BEER HIKING
COLORADO

THE MOST REFRESHING WAY TO DISCOVER COLORFUL COLORADO

D1456537

Beer Hiking Colorado

The most refreshing way to discover colorful Colorado

By: Yitka Winn

ISBN: 978-1-948007-15-3

Published by Helvetiq and VeloPress

Cover Design: Felix Kindelán

Interior Design & Illustrations: Florian Bellon & Janusz Zołociński

Photos: Yitka Winn

Fact-checking & consulting: Brian Metzler

Proofreading: Karin Waldhauser

Printed in Shenzhen, China

First Edition October 2019

All Rights Reserved

Côtes de Montbenon 30, CH-1003 Lausanne

info@helvetiq.ch

4745 Walnut Street, Unit A

Boulder, CO 80301–2587 USA

VeloPress is the leading publisher of books on endurance sports and is a division
of Pocket Outdoor Media. Focused on cycling, triathlon, running, swimming, and
nutrition/diet, VeloPress books help athletes achieve their goals of going faster
and farther. Preview books and contact us at velopress.com.

Distributed in the United States and Canada by Ingram Publisher Services

A Cataloging-in-Publication record for this book is available from the Library of
Congress.

www.helvetiq.com

www.facebook.com/helvetiq

instagram: @helvetiq

19 20 21 / 10 9 8 7 6 5 4 3 2 1

YITKA WINN

BEER HIKING COLORADO

THE MOST REFRESHING WAY TO DISCOVER COLORFUL COLORADO

VELO press®

Boulder, Colorado

HELVETIQ

TABLE OF CONTENTS

1

INTRODUCTION

ABOUT THE AUTHOR

Yitka Winn grew up some 650 miles away from Colorado's mountains, all the way across the neighboring state of Kansas. Every summer, her family road tripped out west, braving the drive across the high plains and wheat fields to spend a couple weeks in a cabin in the Rocky Mountains. Many of her childhood's finest memories are from these trips—craggy, snow-capped peaks, the fragrant scent of sagebrush, the shrieks of yellow-bellied marmots sunning themselves on talus slopes. Hiking was always her family's first mode of choice for exploration; lacing up their boots in pursuit of soaring ridgelines and sparkling alpine lakes.

In later years, when a job opportunity arose in a mountain town in western Colorado, Yitka moved there without hesitation. By then, she'd also developed a healthy appetite for good beer—stouts and porters, especially, though in the process of working on this book, she wound up falling in love with IPAs, too. (Since you asked: Her favorite dark beer featured in this book is the Broken Compass Coconut Porter in Breckenridge. Her favorite IPA is Rare Trait from Denver's Cerebral Brewing.) She also can't resist a good sour.

She has a degree in Creative Writing from Oberlin College, and now works professionally as a writer and editor in the outdoor industry. Her writing has appeared in *Trail Runner*, *Outdoors NW*, Outside Online, REI Co-op Journal, RootsRated, Red Bull Adventure, iRunFar, Gear Institute, and more. If she's not writing or at work, she's probably playing in the mountains.

ABOUT COLORADO

Colorado sits at the epicenter of outdoor exploration in our country. Its mountains burst out of the western flanks of America's flat heartland, giving rise to an exceptional variety of landscapes over the course of just a few hundred miles— 14,000-foot peaks that seem to kiss the sky, lush pine forests, glaciers, Sahara-like sand dunes, canyons rivaling the majesty of the Grand Canyon, rushing rivers, desert rocks and tribal ruins from ancient cliff-dwelling tribes, stands of aspen trees that light up in fiery hues every fall.

If there's anywhere that offers a front seat to the craft beer revolution well underway here in America, it's the state of Colorado. It ranks top in the nation in terms of gross beer production, and currently has more than 400 breweries, with new ones opening up all the time.

Many are situated near (or even at!) trailheads that serve as gateways to the mountains. And, perhaps unsurprisingly, many are owned by outdoor enthusiasts. Their beers are named for peaks and trails and river rapids; their taproom walls are decorated with skis and bicycles and kayak paddles. That iconic, uniquely Coloradoan sense of adventure is baked into every brew.

And, let's be honest—at the end of a long hike in the Colorado sun, there's nothing quite as wonderful as sipping a cold, fine beer.

This guidebook is your ticket to that simple pleasure, to the sublime undertaking of earning one's ales with miles on the trails. Hikes range from short and sweet strolls in the woods to rugged peakbagging missions in the high alpine. The beers span a broad range—juicy IPAs, decadent porters, malty lagers, nutty brown ales and plenty of experimental flavor combinations. (Chai milk stout, anyone? Check out Hideaway Park Brewery in Winter Park.)

Like Colorado's topography itself, there truly is something for everyone.

2

HOW IT WORKS

CHOOSE THE BEER OR THE HIKE

HIKE LOCATION ———————→

NAME OF THE BEER ———————→

MAP ———————————————→

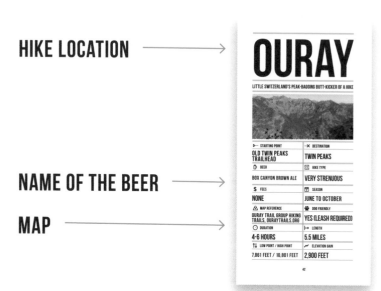

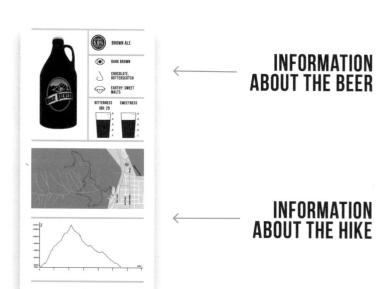

←——— **INFORMATION ABOUT THE BEER**

←——— **INFORMATION ABOUT THE HIKE**

TRAIL AND BEER RATINGS

All ratings in this book—both beer notes and hike difficulties—are subjective. Some hikers might find what I have called a "Moderate" hike to be difficult or even easy. Just like some craft beer drinkers may think that the bitterness or sweetness of a specific beer differ from tasting notes I have described. Please do not consider my ratings as definitive, but as a reference to help you choose the right hike or beer for your taste.

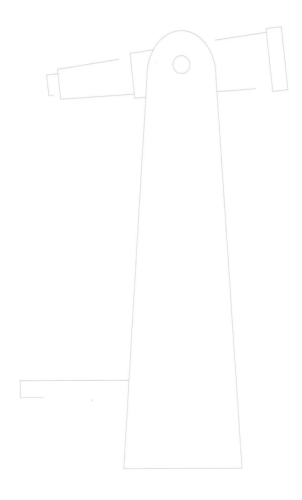

HIKING SAFETY IN COLORADO

Hiking in many parts of Colorado poses some unique risks. In addition to the usual precautions one should take when hiking—notifying someone of your itinerary, carrying the ten essentials, including the knowledge of how to navigate by map and compass, etc.—there are some special factors to be aware of as well, especially if you're visiting from lower-elevation locales.

Many of the hikes in this book venture above tree line, an area that is especially beautiful but also high-risk should thunder clouds roll in. Colorado ranks third in the nation for lightning fatalities. If you can hear thunder, lightning is close enough to strike you; do not venture above tree line in these conditions, or if you're already above it, do everything you can to get below tree line immediately.

Start your hikes early in the day. Afternoons in the months of July and August are often the most dangerous windows of time in Colorado; however, do bear in mind that thunderstorms can strike at any time. Check the weather forecast ahead of time, plan accordingly and, most of all, be willing to modify your plans on the fly for safety reasons.

Additionally, the hikes in this book generally occur at higher altitudes than a vast majority of hikes elsewhere in the continental U.S. Be sure to bring a hat, sunscreen, sunglasses and plenty of water, especially on hikes that are more exposed. It is easier to get sunburned at altitude than at sea level.

For high-elevation hikes, pay attention for any signs of altitude sickness, such as shortness of breath, nausea, fatigue or headaches. Especially if you're coming from sea level, exerting yourself above 8,000 feet can trigger symptoms—and even more intensely so once you get above 12,000 feet. For mild symptoms, drink plenty of water, slow your pace and take plenty of breaks. For more severe symptoms (such as an unsteady gait, intensifying nausea, vomiting or shortness of breath even when resting), get yourself down to lower elevation as soon as possible and, if symptoms persist, seek medical attention.

In the winter and spring, avalanches are also a risk factor in many parts of Colorado—most notably for backcountry skiers or snowboarders, mountaineers and snowshoers. Of the hikes in this book, the ones in the backcountry are written primarily with summer and/or occasionally autumn use in mind. Do not attempt any routes when large swaths of snow are still present unless you've done your homework ahead of time on avalanche safety awareness and preparation.

The mountains of Colorado are home to many kinds of majestic wildlife, the vast majority of which pose no risk whatsoever to humans. However, it's wise to read up on wildlife encounters, safety precautions, and how to react should you come face to face with any critters that can, in rare situations, pose a danger to humans—namely, rattlesnakes, black bears, moose and mountain lions. Arm yourself with knowledge.

Be sure to check your skin for ticks after hiking.

Lastly, a word on beer-imbibing safety as well: be aware that, particularly after a hard hike at altitude, your body is likely to already be dehydrated, which can make alcohol feel as though it hits you much harder and faster. Drink plenty of water and consume ample food during and after your hikes. Don't expect your alcohol tolerance to feel equivalent to what it may be at lower elevations.

The majority of the trailheads and breweries in this book require a vehicle to access. Drink responsibly, and never drink and drive. Have a designated driver, or wait at least 45 minutes per beer you enjoy for the alcohol to clear your system before getting behind the wheel of a vehicle. Otherwise, plan to get a growler, six-pack or cowboy can to go, and enjoy your beer responsibly back at your home, campsite, cabin, RV or hotel.

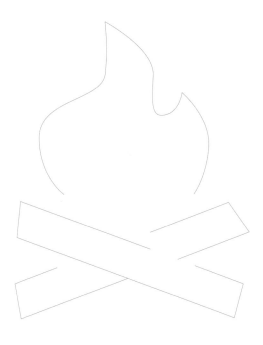

TRAIL ETIQUETTE

Though this is a book on hiking and beer drinking, not on how to be a decent human when you venture into the wild, I'd be remiss if I didn't provide a few tips on trail etiquette.

Colorado is a popular place for outdoor enthusiasts. This book includes a mix of some lesser known trails and more popular (i.e., crowded) hikes. Especially if you're hiking on weekends, you're unlikely to have many of the trails in this book to yourself. Be respectful of other trail users and of the environment by following some simple rules:

- Stay on trail—don't cut switchbacks as you risk trampling delicate flora and encouraging erosion.

- Limit your group size—especially in designated wilderness areas, where groups of 12 or larger are prohibited.

- Keep your noise level to a respectful volume—leave your Bluetooth speakers at home, please.

- Observe Leave No Trace (LNT) principles—respect wildlife, pack out your trash, educate yourself on how to handle bathroom needs in the backcountry. Learn more at LNT.org.

- Yield to uphill hikers, who have the right of way.

- On equestrian trails, yield to horses, giving them plenty of space to get by you.

- Stay aware of your surroundings. (e.g., Do not hike with noise-cancelling headphones.)

Finally, many of the hikes in this book (though not all) are dog friendly. If you like to hike with a four-legged companion, please do so with the utmost respect for other trail users as well as the land manager's rules, including leash regulations, which are often in place to help protect wildlife and fragile ecosystems. Pick up your dog's waste—and please don't leave plastic baggies of poop on the side of the trail to pick up on your way back.

A NOTE ON THE BEERS

While I've tried to primarily recommend beers that can be enjoyed at the brewery year-round, there are some exceptions. Some breweries are more experimental and put out new beers on constant rotation, meaning there isn't a set of core beers always on tap. The craft beer scene in the area is a dynamic one and new beers sometimes take the place of old favorites to keep up with the changing landscape.

I chose beers based on brewery recommendations, community reviews, and my own personal enjoyment. The highlighted beers are certainly not the only ones worth trying at any given brewery. I fully encourage you to sample any beer that catches your fancy; many breweries offer sampler "flights" to encourage just this.

Consider this book your sampler flight of hikes and brews in Colorado. There are many more incredible trails, breweries and beers to discover on your own. Let this book be your springboard for a lifetime of beer-hiking explorations in the Centennial State and beyond.

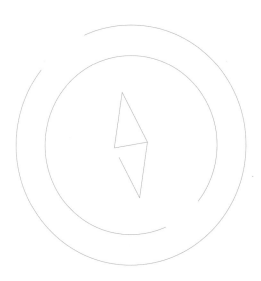

BEFORE SETTING OFF

TEN ESSENTIALS

Even if you're just heading out on a day hike, you should always be prepared. Here's a list of the ten essentials you should take with you on the trail.

- Navigation; Maps in this book are for reference only. The most reliable form of navigation is a topographic paper map and compass. Batteries die—don't rely solely on your phone!
- Hydration; Water is essential when you're hiking. Make sure you bring plenty along to keep yourself hydrated and ward off the potential for dehydration and heat stroke. Don't worry—no matter how much water you drink, you'll still be thirsty enough for a beer!
- Nutrition; Bring extra food along just in case you're on the trail for longer than you expect. Foods that are high in protein and can be stored for a while can be kept in your pack for emergencies.
- Rain Gear and Insulation; The weather can change quickly! Rain gear is especially important; afternoon thunderstorms are common in the summer in Colorado. Bring an extra layer of clothing as temperatures fluctuate in the backcountry.
- Fire Starter; In case of emergency, a water-tight container of water-proof matches is invaluable to a lost hiker for warmth and cooking.

- First Aid; From minor cuts and scrapes, to more major injuries, you need to be prepared.
- Tools; From a pocket knife to a multi-tool, these small tools take on big jobs in the backcountry. And never underestimate the power of duct tape!
- Illumination; Whether the hike took longer than you anticipated or you really wanted to catch that sunset from the top, you could get caught in the dark. Bring a flashlight or headlamp so you'll be able to see the trail.
- Sun Protection; Sunglasses, hats, and sunscreen keep you from getting sunburned and protect your eyes from the bright glare.
- Shelter; As simple as a space blanket or tarp—it could keep you dry in inclement weather or unexpected overnight on the trail.

HIKING SEASON

Seasons are always a factor when hiking in Colorado. High country trails may not be accessible between late fall and spring due to snow levels. Always check with the land manager (listed for each hike) for current trail conditions before setting out. Local tourism offices can be a good resource as well.

WEATHER

Because conditions can change in an instant—especially in the mountains—it's best to always be prepared for wind, rain, snow, and anything else Mother Nature might throw your way. For accurate forecasts, check the following site:

- weather.gov

ADDITIONAL RESOURCES

For additional hiking and brewery information, I suggest the following sites:

HIKING

- gohikecolorado.com
- hikingproject.com
- alltrails.com

BREWERIES

- coloradocraftbrews.com
- coloradobrewerylist.com

MAP

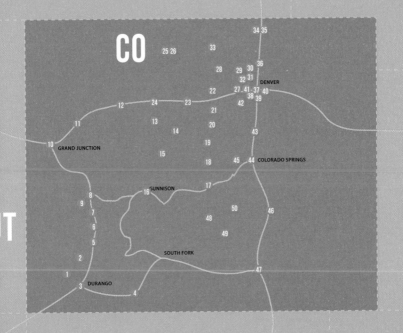

3

TRAILS

CORTEZ

A PEACEFUL DESERT TOUR THROUGH CANYON OF THE ANCIENTS

▷⋯ STARTING POINT	⋯✗ DESTINATION
SAND CANYON TRAILHEAD	**ASSORTED RUINS**
🍺 BEER	🔳 HIKE TYPE
SCHNORZENBOOMER AMBER DOPPELBOCK	**EASY-MODERATE**
$ FEES	📅 SEASON
NO	**MARCH TO NOVEMBER**
⛰ MAP REFERENCE	🐾 DOG FRIENDLY
CANYONS OF THE ANCIENTS ON BLM.GOV	**YES (LEASH REQUIRED)**
🕐 DURATION	↦ LENGTH
2-3 HOURS	**5.5 MILES**
↑↓ LOW POINT / HIGH POINT	〰 ELEVATION GAIN
5,456 FEET / 5,909 FEET	**550 FEET**

ALCOHOL 8.5% CONTENT

AMBER DOPPELBOCK

 DARK AMBER

 BUTTERSCOTCH

 MALT, CHERRIES

BITTERNESS
IBU: UNLISTED

SWEETNESS

5
4
3
2
1

5
4
3
2
1

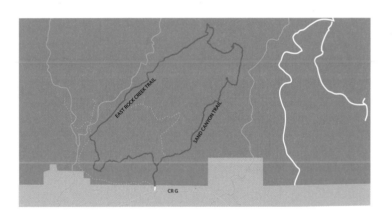

EAST ROCK CREEK TRAIL

SAND CANYON TRAIL

CR G

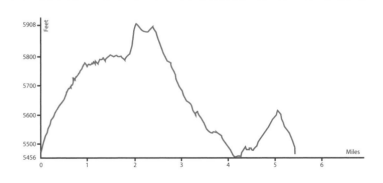

Feet

5908
5800
5700
5600
5500
5456

0   1   2   3   4   5   6

Miles

DESCRIPTION OF THE ROUTE

Take a peaceful desert tour through this outdoor museum of ancient ruins before stopping in nearby Cortez for a "mesa cerveza."

Situated on the southern flank of the massive Canyons of the Ancients National Monument, this scenic loop is peppered with ancient cliff dwellings, Puebloan ruins, caves, sandstone bluffs, and sprawling canyons. It's the closest you can get to the otherworldly landscapes of Utah's famous national parks (think Bryce, Zion, and Arches) without actually crossing the state border. The Monument itself contains more than 6,000 archaeological sites and artifacts dating back a millennium to the ancient Anasazi and Puebloan people, many of whom once farmed in the valley.

The Sand Canyon trailhead is a bit of a haul from downtown Cortez, but don't let this discourage you. It's an easy, quick paved road out to the trailhead at a small slickrock parking area.

Start up the slickrock, following the orange diamonds and large cairn to a more obvious trail. You'll pass through trees on a trail with both footprints and bike tracks. (Mountain biking and horseback riding are also permitted in the Monument, so do keep an eye out.) Stay left, following the wooden posts over the slickrock.

Along the way, you'll pass small trail spurs marked with white diamonds. These frequently lead to ruins or artifacts, so be on the lookout for these and don't skip the opportunities for short, worthwhile detours before rejoining the main trail. One of the first you'll come to is the Castle Rock Pueblo.

Endless trail options exist here. If you carry a map, which you can pick up at the Anasazi Heritage Center in nearby Dolores, you can shorten or extend this loop as desired. (Major trail intersections along the loop have "You Are Here" signs with maps, too.) Carry plenty of water with you, and don't forget your hat, sunscreen, and sunglasses. Avoid this trail at the height of summer, for temperatures can be scorching and there is virtually no shade.

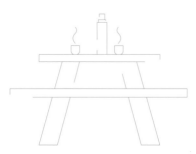

TURN BY TURN DIRECTIONS:

1. Follow the orange diamonds. White diamonds often lead to short trail spurs to ruins.
2. At 1.8 miles, go left at the junction.
3. At 2.5, go left to take the East Rock Creek Trail.
4. At 3.5, stay right; at 4.1, go left, following the blue blaze.
5. At 4.5, follow wooden posts over the slickrock.
6. At 5.25, go right to complete your loop.

FIND THE TRAILHEAD

Head south of Cortez on Highway 491 for 2.5 miles. Make a right on County Road G (signs for Hovenweep National Monument), drive 12.5 miles and find the small Sand Canyon trailhead on your right. Park on the slickrock.

MAIN STREET BREWERY

Cleverly dubbing their brews "Mesa Cerveza," the Main Street Brewery in downtown Cortez offers up an extensive line of in-house beers. Many are European-style brews, ranging from German pilsners and doppelbocks to Scottish ales and Irish stouts. For a truly one-of-a-kind experience, order one of their signature blends. These two-beer combinations are mixed at the tap, such as the "Boomerang"—a super-sweet blend of the Schnorzenboomer and their Honey Raspberry Wheat.

CONTACT INFORMATION
Canyons of the Ancients and Anasazi
Heritage Center,
27501 Highway 184,
Dolores, CO 81323;
970-882-5600

BREWERY/RESTAURANT
Main Street Brewery
21 E Main Street,
Cortez, CO 81321
Miles from trailhead: 16

DOLORES

HIKE ALONG THE BLUFFS ABOVE MCPHEE RESERVOIR

▷··· STARTING POINT	···✕ DESTINATION
4TH STREET AND ABEYTA DRIVE	**MCPHEE OVERLOOK**
🍺 BEER	HIKE TYPE
ESB	**EASY-MODERATE**
$ FEES	📅 SEASON
NONE	**MAY TO OCTOBER**
⌂ MAP REFERENCE	🐾 DOG FRIENDLY
TOWN OF DOLORES MCPHEE OVERLOOK TRAIL PROJECT SITE MAP AT CPW.STATE.CO.US	**YES**
⌚ DURATION	↦ LENGTH
2 HOURS	**3.7 MILES**
↑↓ LOW POINT / HIGH POINT	〰 ELEVATION GAIN
6,909 FEET / 7,057 FEET	**140 FEET**

 ESB

 HAZY, DARK COPPER

 MAPLE AND HERBS

MALTY, ROASTY, COCOA

BITTERNESS
IBU: 24

SWEETNESS

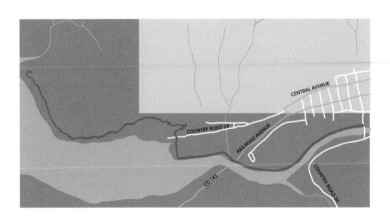

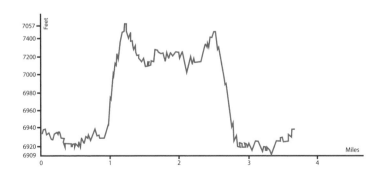

DESCRIPTION OF THE ROUTE

Take an easy spin through southwest Colorado's high desert before pulling up a stool at the neighborhood watering hole for pizza and a brewski.

Developed recently in large part by community volunteers, the lovely 7-mile McPhee Overlook Trail connects the town of Dolores to the popular Boggy Draw trail system, which provides some 35 miles of non-motorized singletrack for mountain bikers, hikers, and trail runners alike. Proffering bird's-eye views of the expansive McPhee Reservoir, the trail winds along the edges of bluffs, through shrubby gardens and past impressive rock formations. Watch out for little lizards darting along the edges of the trail.

For an easy jaunt, you can literally start and finish from the doorstep of the brewery (or anywhere else along the main drag in downtown Dolores) by making your way to the south end of 4th Street. There, pick up the Dolores River Trail, a flat, gravel path paralleling the river. Along the way, pass fishing holes and an epic playground for kiddos.

At just under a mile, you'll reach the parking area for McPhee Overlook Trail. Follow the short series of switchbacks up through the woods and toward the rocky bluffs above. Once the initial climb is over, the grade mellows as the singletrack winds along the edges of the bluffs overlooking the sparkling reservoir. Pinyon pine and juniper trees dot both sides of the trail.

The beauty of this hike is that you can go out for as long (or as short) as you like. At 1.8 miles, there's a nice overlook before the trail turns briefly away from the reservoir to dip into a shrubby cove. Beyond that, if you hike on for all 7 miles (14 miles total as an out-and-back), you'll eventually reach the House Creek Forest Service campground before connecting to the Boggy Draw trail system for non-motorized recreation.

Relative to many other hikes in this part of the state, this hike's lower elevation and desert-like exposure make it a good pick for late spring or early fall hiking. Be aware that while the Dolores River Trail is open year-round, seasonal wildlife closures restrict access to the McPhee Overlook Trail between December 1 and May 1 every year.

TURN BY TURN DIRECTIONS:

1. Start on the flat, gravel path along the river.
2. At 0.9 miles, reach the parking area for McPhee Overlook Trail. Cross the parking area and venture up past the trailhead sign into the woods.
3. Hike as far out as you like before turning around.

FIND THE TRAILHEAD

From downtown Dolores, head south on 4th Street and pick up the Dolores River Trail on your right-hand side, just before the river.

DOLORES RIVER BREWERY

Just down the road from the start and finish of this lollipop-loop hike to the reservoir, the Dolores River Brewery opens up at 4 p.m. most days for locals and visitors alike to gather, imbibe, and partake in its delicious wood-fired pizza. It has a rustic, cozy neighborhood feel to it. Beers rotate seasonally, but this easy-drinking English-style pale ale ESB is a perennial favorite.

CONTACT INFORMATION
U.S. Forest Service,
Dolores Ranger District,
29211 Highway 184,
Dolores, CO 81323;
970-882-7296

BREWERY/RESTAURANT
Dolores River Brewery
Address: 100 4th St, Dolores, CO 81323
970-882-4677
Miles from trailhead: 0.2

DURANGO

A SHORT, STEEP LOLLIPOP-LOOP HIKE AND RIDGE SCRAMBLE

▷⋯ STARTING POINT	⋯✗ DESTINATION
WEST END OF LEYDEN STREET	**HOGSBACK**
🍺 BEER	HIKE TYPE
COLORADO KÖLSCH	**STRENUOUS**
$ FEES	SEASON
NONE	**MARCH TO NOVEMBER**
⌂ MAP REFERENCE	🐾 DOG FRIENDLY
OVEREND MOUNTAIN PARK MAP ON DURANGOGOV.ORG	**YES**
⏱ DURATION	↦ LENGTH
1.5-2 HOURS	**2.5 MILES**
↑↓ LOW POINT / HIGH POINT	〰 ELEVATION GAIN
6,686 FEET / 7,493 FEET	**780 FEET**

 KÖLSCH

 PALE STRAW

 BISCUITY; HINTS OF PEAR

 CRISP, DRY, FIZZY

BITTERNESS
IBU: 16

SWEETNESS

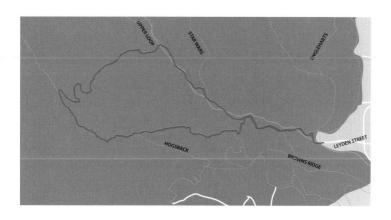

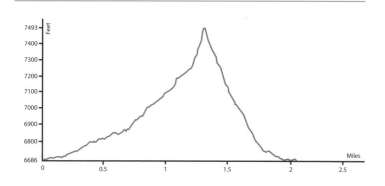

DESCRIPTION OF THE ROUTE

A short, steep lollipop-loop hike and ridge scramble, rewarded with a crisp Colorado Kölsch to quench your thirst.

Hogsback Ridge looms over downtown Durango like a sleeping giant. Snaking up the spine of Hogsback is a narrow, steep trail visible from just about anywhere in town. The foothills along its flanks comprise Overend Mountain Park, a city-owned treasure of singletrack for hiking, mountain biking, and trail running. Though hiking opportunities abound in all directions, standing atop the iconic Hogsback is a veritable rite of passage for locals.

Numerous trailheads exist for Overend Mountain Park, but the easiest starting point for a short jaunt on Hogsback is the trailhead tucked into a residential neighborhood at the west end of Leyden Street. From here, trail maps are posted on signs at nearly every intersection, so it's easy to navigate as you go.

This loop option takes you up the backside of Hogsback first, following the popular Perins Gulch trail. (Watch out for mountain bikers, as the lower trails in Overend are a popular haunt for a variety of trail users.) Follow the main Perins Gulch trail through ponderosa and pinyon pines, sandy washes, and scrubby vegetation.

At 0.6 miles, you'll turn left to get on the Hogsback trail, which switchbacks gently up out of the woods, eventually popping out on the backside ridgeline of Hogsback. Enjoy the up-close view of another Durango icon, nearby Perins Peak. On the other side, you'll have a terrific view of bustling downtown Durango and the Animas River.

If you have a fear of heights or are wearing shoes that lack good traction, this is a good place to turn around and go back the way you came. The trail that lies ahead is never cliffy, but it is narrow and drops off dramatically on both sides. After veering left, you'll begin the absurdly steep hike (or scramble, in some places) to follow the spine and attain the summit. You'll know you're there when you reach the large, flat boulder affording 360-degree views.

If you know what you're looking for, you can glimpse the back deck of the Steamworks Brewing Co. from here, where a crisp, refreshing Kölsch awaits you. Facing Durango, the trail down is to the left of the boulder. This trail can feel like you're plummeting down a roller coaster, stomach drops and all. Hold it together and pick your way down the steepest parts carefully until you hit the intersection with 8 Bells. Bear left to hook back in with the Perins Gulch trail and complete your lollipop loop.

TURN BY TURN DIRECTIONS:

1. From the end of Leyden Street, follow Perins Gulch Trail.
2. At 0.4 miles, veer left to stay on Perins Gulch.
3. At 0.6, make another left onto the Hogsback trail and cross a footbridge.
4. At 1.1, pop up on the ridgeline. Go left until you reach the summit at 1.3 miles.
5. Descend steeply on the frontside Hogsback trail.
6. When you reach 8 Bells Trail, make a left to hook back in to Perins Gulch Trail and return to the trailhead.

FIND THE TRAILHEAD

From downtown Durango, drive west on 9th Street. Turn right on Roosa Avenue, left on El Paso Street, right on Forest Avenue, then left on Leyden Street. The trailhead is at the end of this street. Park on the street.

STEAMWORKS BREWING COMPANY

Durango is affectionately called the "Napa Valley of Beer," as many of its breweries are among the oldest in the state. Brewing since 1996, Steamworks sports a warehouse-like atmosphere and sizable outdoor deck with a straight-shot view of Hogsback. As you quench the thirst you worked up hiking, you can enjoy gazing at exactly what you accomplished with a crisp, German-style Kölsch in hand, a perennial award winner at beer festivals around the world.

CONTACT INFORMATION
City of Durango Parks and Recreation,
970-375-7321

BREWERY/RESTAURANT
Steamworks Brewing Company
801 E 2nd Avenue,
Durango, CO 81301
970-259-9200
Miles from trailhead: 1.4

PAGOSA SPRINGS

VENTURE INTO AN OUTDOOR CATHEDRAL OF WATERFALLS

▷⋯ STARTING POINT	⋯✗ DESTINATION
FOURMILE FALLS TRAILHEAD	**FOURMILE FALLS**
🍺 BEER	▦ HIKE TYPE
EL DUENDE VERDE CHILE ALE	**EASY/MODERATE**
$ FEES	📅 SEASON
NONE	**MAY TO SEPTEMBER**
⌂ MAP REFERENCE	🐾 DOG FRIENDLY
TRAILS ILLUSTRATED 145: PAGOSA SPRINGS, BAYFIELD	**YES**
🕐 DURATION	↦ LENGTH
3-4 HOURS	**6.2 MILES**
↑↓ LOW POINT / HIGH POINT	∿ ELEVATION GAIN
8,996 FEET / 9,895 FEET	**1,080 FEET**

ALCOHOL 5% CONTENT

PALE ALE

 DARK GOLDEN, SLIGHTLY HAZY

 GREEN CHILE PEPPERS, BREAD

LIGHT, DRY; SLIGHT KICK

BITTERNESS
IBU: 34

SWEETNESS

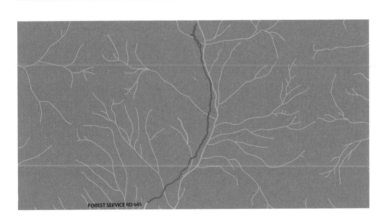

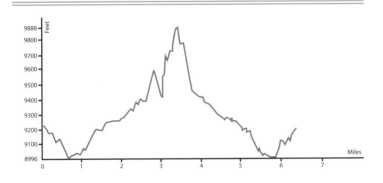

FOREST SERVICE RD 645

DESCRIPTION OF THE ROUTE

Spice up your day with waterfalls and a green-chile-spiked pale ale in this off-the-beaten-path gem of a mountain town.

Surrounded by three million acres of wilderness areas and national forests, Pagosa Springs is the perfect base camp for adventure. It boasts multiple breweries for tasting connoisseurs, and is jam packed with mineral hot springs pools to soak weary muscles after a long day on the trail. In the winter, it's also a hit with powder hounds, thanks to its proximity (24 miles) to Wolf Creek Ski Area, which frequently receives more snow annually than any other ski area in the state of Colorado.

Nestled in the Weminuche Wilderness Area, this classic Pagosa hike is a little ways outside of town—but it's well worth the drive to the trailhead. The hike's namesake falls sit in the belly of a steep valley, flanked on all sides with prominent rocky bluffs and hoodoo-like outcroppings. You'll feel as though you're venturing into an immense outdoor cathedral.

The rocky trail begins slightly downhill, crossing some small creeks (larger in the spring). After a half-mile, you'll pass the Weminuche Wilderness Area boundary and catch your first glimpses of the looming peaks surrounding you.

Unfortunately, this area has been hit hard in recent years by beetle kill, so the pine trees are not what they used to be. Nevertheless, wildflowers abound in the summer, and prolific copses of aspen light up the valley at the height of fall.

The trail is easy to follow. Before you reach Fourmile Falls, you'll first come to the stream at the foot of the almost equally impressive Falls Creek Falls. They spill from a tall cliff to your left, with peak runoff in the late spring as Colorado's snowpack melts out of the high alpine. You'll have to crane your neck to get a good view here!

After you've had your fill, cross the creek and continue on to Fourmile Falls. You'll get an excellent view of them from the trail around 3 miles in, at which point you'll have just a couple more minutes to clamber your way up the trail to the top of them. A peaceful lunch spot awaits you next to the rushing river above the falls.

TURN BY TURN DIRECTIONS:

1. Follow the main trail; the sign says Fourmile Stock Drive Tr. 569.
2. At 0.5 miles, enter Weminuche Wilderness Area.
3. At 2.7, the trail appears to fork. Go left/up to reach creek at the bottom of Falls Creek Falls.
4. After crossing the creek, come to an intersection. Go left to get closer to the falls or go right to stay on the main trail to Fourmile Falls at 3.1 miles.

FIND THE TRAILHEAD

From downtown Pagosa Springs, head west on Pagosa Street. Turn right onto Lewis Street, then left onto N 5th Street and continue on Four Mile Road (County Road 400) for 7.5 miles. Make a right at the fork and continue another 4.5 miles on Four Mile Road, which becomes Forest Service Road #645, to reach the trailhead.

RIFF RAFF BREWING COMPANY

Don't miss "hoppy hour" at this friendly neighborhood joint on the main drag through historic downtown Pagosa. They pour their own flagships brews, guest beers, and seasonal picks such as a black cherry porter, plum wheat, and "spruce juice," brewed with locally harvested spruce tips. An environmentally forward-thinking brewery focused on "keeping the earth hoppy," Riff Raff is heated by geothermal heat from the world's deepest hot spring.

CONTACT INFORMATION
U.S. Forest Service,
Pagosa Ranger District,
180 Pagosa Street,
Pagosa Springs, CO 81147;
970-264-2268

BREWERY/RESTAURANT
Riff Raff Brewing Company
274 Pagosa Street,
Pagosa Springs, CO 81147
970-264-4677
Miles from trailhead: 13.6

SILVERTON

A STUNNING, HIGH-ALTITUDE LOOP THROUGH ALPINE TUNDRA

▷··· STARTING POINT	···✕ DESTINATION
CUNNINGHAM GULCH	**HIGHLAND MARY LAKES**
🍺 BEER	🔁 HIKE TYPE
PRIDE OF THE WEST PORTER	**MODERATE-STRENUOUS**
$ FEES	📅 SEASON
NONE	**JULY TO SEPTEMBER**
⛰ MAP REFERENCE	🐾 DOG FRIENDLY
LATITUDE 40° MAPS TELLURIDE-SILVERTON-OURAY	**YES (LEASH REQUIRED)**
🕐 DURATION	↦ LENGTH
5-7 HOURS	**7.7 MILES**
↕ LOW POINT / HIGH POINT	〜 ELEVATION GAIN
10,820 FEET / 12,612 FEET	**1,900 FEET**

 PORTER

 **DARK BROWN;
CARAMEL HUE**

COFFEE, TOFFEE

 **CHOCOLATE,
ROASTED MALT**

**BITTERNESS
IBU: 20**

SWEETNESS

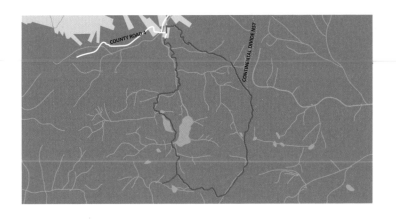

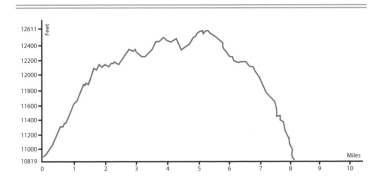

DESCRIPTION OF THE ROUTE

Celebrate this stunning, high-altitude loop through alpine tundra with a rich porter in Silverton's historic district.

"Pride of the West" is a term to describe many things—a now-defunct dive bar in the historically rowdy miners' town of Silverton, the sweet porter that awaits you at the end of this hike at Avalanche Brewing Company, or the San Juan Mountains themselves—arguably one of the most impressive mountain ranges in the country.

The trailhead for Highland Mary Lakes takes a little more work than others to reach, which is perhaps why this unbelievable, high-alpine loop hike remains less crowded than other hikes in the San Juans. Along your hike, though, you'll pass everything from pristine waterfalls to meadows blanketed in wildflowers to a number of shimmering lakes.

The steepest part occurs right off the bat as you make your way up the gulch along a cascading creek on your way to the Weminuche Wilderness Area boundary. Shortly after, there are some confusing social trails. Stay on the main route by going right up over some big rocks, then left through the woods and across another a creek.

At 1 mile, a simple wooden "Trail" sign offers an ambiguous arrow. Make a hard right here across the creek. Follow cairns. In another half-mile, the trail forks. Either way will do, as they'll rejoin shortly. The left way along the creek is easiest. The right way through the talus garden is more circuitous and scrambly. Either way, soon you'll arrive at the first of the three Highland Mary Lakes.

Past the third lake, follow the wooden posts up to the ridgeline saddle. From here, you'll be able to glimpse the iconic jagged teeth of the Grenadier range's Arrow and Vestal peaks ahead of you, and Verde Lakes down to your right—a short, optional out-and-back addition. Whether or not you take this side trip, your loop hike continues by veering left (when facing the Grenadiers) at the saddle, following the wooden posts farther up the ridge.

After the Continental Divide Trail (CDT) feeds in to the trail from your right, you may see some CDT or Colorado Trail thru-hikers; your route shares a short stretch with these celebrated long-distance trails. In a little over a mile, though, you'll stay straight/left at the CDT turn-off sign to make your way back to your trailhead. In another half-mile at the signed trail fork, stay straight/right to take a different path down to the Continental Divide Trailhead, just a scoot down the road from the trailhead where you started your day.

A few final words of advice: start early in the day and do not attempt this loop if thunderstorms are forecasted, as it lies almost entirely above tree line. Also, many intersections are unmarked, so bring a topo map.

TURN BY TURN DIRECTIONS:

1. At the well-signed intersection at 0.25 miles, stay right on the main trail.
2. At 1.5, the trail forks. Stay left for the easier, more direct route.
3. Stay right around first small lake. Stay left around second, larger lake. Stay right around third. Follow wooden posts up to the saddle.
4. When you reach the saddle, either make a quick (0.5-mile) out-and-back trip to Verde Lakes to your right, or veer left to continue on the loop, following wooden posts.
5. At 5.7 (or 6.2 if you went to Verde Lakes), stay straight/left, passing the CDT sign.
6. At 6.9 (or 7.4 if you went to Verde Lakes), stay straight/right for new scenery that still loops you back to your starting point.

FIND THE TRAILHEAD

Drive northwest out of Silverton on County Road 2, which turns to gravel in 2 miles. In 2 more miles, make a right on County Road 4 and follow it to the trailhead in 5 more miles. The final mile is a bit rough but still doable with a 2WD vehicle.

AVALANCHE BREWING COMPANY

Avalanche Brewing Company is situated in downtown Silverton on Blair Street, an old dirt road notorious for its rowdy history of saloons, bordellos, and dance halls. Housed inside a quaint, kitschy house reminiscent of a ski-bum shack, Avalanche serves up its own crafted "elevated ales" alongside breakfast, lunch, and artisan thick-crust pizza. Grab a seat on the porch for a front-seat view of Kendall Mountain.

CONTACT INFORMATION
U.S. Forest Service,
Silverton Field Station,
1468 Greene Street,
Silverton, CO 81433;
970-387-5530

BREWERY/RESTAURANT
Avalanche Brewing Company
1067 Blair Street,
Silverton, CO 81433
970-387-5282
Miles from trailhead: 9.2

OURAY

LITTLE SWITZERLAND'S PEAK-BAGGING BUTT-KICKER OF A HIKE

▷⋯ STARTING POINT	⋯✕ DESTINATION
OLD TWIN PEAKS TRAILHEAD	**TWIN PEAKS**
🍺 BEER	HIKE TYPE
BOX CANYON BROWN ALE	**VERY STRENUOUS**
$ FEES	SEASON
NONE	**JUNE TO OCTOBER**
⌖ MAP REFERENCE	🐾 DOG FRIENDLY
OURAY TRAIL GROUP HIKING TRAILS, OURAYTRAILS.ORG	**YES (LEASH REQUIRED)**
🕐 DURATION	↦ LENGTH
4-6 HOURS	**5.5 MILES**
↑↓ LOW POINT / HIGH POINT	〰 ELEVATION GAIN
7,861 FEET / 10,801 FEET	**2,900 FEET**

 ALCOHOL 5.6% CONTENT

BROWN ALE

 DARK BROWN

 CHOCOLATE, BUTTERSCOTCH

EARTHY; SWEET MALTS

BITTERNESS
IBU: 29

SWEETNESS

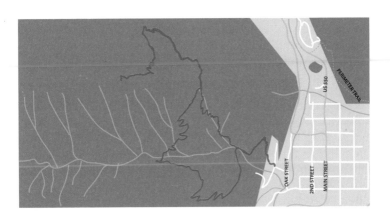

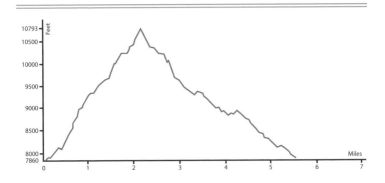

DESCRIPTION OF THE ROUTE

This peakbagging butt-kicker of a hike offers a challenging, respectable preamble to your Box Canyon Brown Ale on the Ouray Brewery's sunny rooftop deck.

Nicknamed "Little Switzerland" for good reason, Ouray is an old mining town surrounded by steep cliffs, peaks, canyons, and waterfalls. In the winter, a picturesque gorge on the edge of town offers some 200 of the world's best ice and mixed climbing routes. In the summer, buttery singletrack snaking up into the surrounding mountains and canyons beckons hikers.

The rocky summits of Twin Peaks are visible from just about anywhere in town, including the sun-drenched rooftop deck of the Ouray Brewery—a can't-miss stop for locals and tourists alike on Ouray's main drag.

From the Old Twin Peaks trailhead, start up the gravel road before joining the singletrack that rapidly steepens, switchbacking up through woods, talus, and pink-tinged rock cliffs. Within a quarter mile, you'll have great views of the town below, the ice park canyon walls and, across the valley, Ouray's infamous natural "Amphitheater."

After you pass the fenced-off Henns' Overlook, the trail wraps right and you'll begin to hear the rushing rapids of Oak Creek. Look up and you'll see Twin Peaks towering above in the sky. Soon, just shy of the large bridge over Oak Creek, look to your right for two signs nailed to a tree—one that says "Old Twin Peaks Trail," one that reads (forebodingly, but accurately), "Caution: steep and difficult trail." Veer off the well-established Perimeter Trail here and up onto the narrow, rocky Old Twin Peaks trail.

From here, the trail is a butt-kicker—literally a staircase of logs and cribbing in places, passing avalanche chutes, mining ruins, lush forest, and aspen stands. At just shy of 2 miles, you'll pop out on the ridgeline with expansive views of the Camp Bird valley and the highway up to Red Mountain Pass. To your left is a scenic overlook at Sister Peak. To your right, the trail continues—a steep, loose calf-burner—up to the scrambly Twin Peaks summit. Enjoy soaring views of the Sneffels and Cimarron mountain ranges, and—if you know where to look—the rooftop deck of the Ouray Brewery where you'll soon be resting your weary legs and enjoying a cold one.

You can either return the same way you came up or, for a change of scenery and a longer but slightly more forgivingly graded descent, take the Oak Creek Trail on your way down. To do so, hang a right at the signed junction at 3.3 miles, following signs for Oak Creek. At 4 miles, cross the creek, then go left and down at the signed trail junction shortly after. (Note: at peak spring runoff in May or early June, this creek may not be passable.) At 5 miles, hang a left on the Perimeter Trail to cross the bridge over Oak Creek and return to town on the same trail you first came up.

TURN BY TURN DIRECTIONS:

1. Start up the gravel road before quickly getting on singletrack.
2. At 0.4 miles, about 50 feet shy of the bridge, veer off of the Perimeter Trail and up onto the narrow, rocky Old Twin Peaks trail on your right.
3. At 1 mile, make a left, then a right at the well-signed trail junction, following signs for Twin Peaks.
4. At 2.2, reach the summit.
5. Complete a loop by following signs for Oak Creek on your way down, then making a left on the Perimeter Trail.

FIND THE TRAILHEAD

From downtown Ouray, walk or drive west on 7th Avenue to cross the Uncompahgre River. Go right on Oak Street, then left on Queen Street until it meets Pinecrest Drive. A few parking spots exist at the trailhead.

OURAY BREWERY

With a rooftop terrace surrounded on all sides by steep canyons and dramatic mountains—including the visible dual summits of Twin Peaks— the Ouray Brewery is *the* spot to be on sunny afternoon in the San Juan Mountains. Its old-timey saloon vibe is complemented by the comfiest swinging barstools you've ever sat on and a diverse lineup of beers—many named for old local mining roads, trails, and other relics of a bygone era.

CONTACT INFORMATION
U.S. Forest Service,
Ouray Ranger District,
2505 S. Townsend,
Montrose, CO 81401;
970-240-5300

BREWERY/RESTAURANT
Ouray Brewery
607 Main Street,
Ouray, CO 81427
970-325-7388
Miles from trailhead: 0.5

RIDGWAY

SUMMIT THE *GUNSMOKE*-THEMED LOCAL FOOTHILL

▷⋯ STARTING POINT	⋯✕ DESTINATION
NORTH END OF RAILROAD ST.	**BOOT HILL SUMMIT**
🍺 BEER	HIKE TYPE
IRISH RED ALE	**EASY-MODERATE**
$ FEES	SEASON
NONE	**APRIL TO OCTOBER**
MAP REFERENCE	🐾 DOG FRIENDLY
DENNISWEAVERPARK.COM	**YES**
⏱ DURATION	↦ LENGTH
1.5-2 HOURS	**3.75 MILES**
↑↓ LOW POINT / HIGH POINT	〰 ELEVATION GAIN
6,952 FEET / 7,172 FEET	**220 FEET**

 RED ALE

 CLEAR RED/AMBER

CARAMEL

MALTY, EARTHY

BITTERNESS
IBU: 25

SWEETNESS

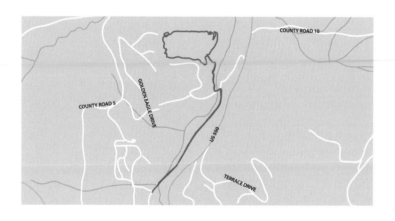

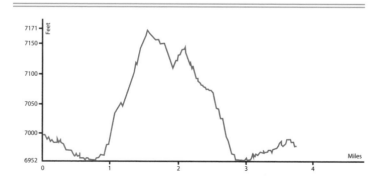

DESCRIPTION OF THE ROUTE

Get a taste for the Wild West with this hike through a wildlife reserve and to the summit of *Gunsmoke*-themed local foothill.

The tiny town of Ridgway carries an immense amount of history, having once served as home of the Ute Indians, then in the 1890s as the railroad gateway to the nearby mining districts of Ouray, Telluride, Rico, and Durango. In the 1960s, the Hollywood movies *How the West was Won* and *True Grit* were both filmed here, and to this date, the town retains its moniker of "Gateway to the San Juans"—the majestic, craggy peaks that loom over Ridgway's southern horizon.

For some 14 years, Dennis Weaver, a longtime environmentalist and famous actor who played Chester Goode in the old TV western, *Gunsmoke*, called Ridgway home before passing away in 2006. Today, a beautiful, 60-acre park and permanent wildlife reserve dedicated in his memory exist on the northern flanks of Ridgway, sandwiched between downtown and the expansive Ridgway State Park (where further trails abound).

A paved bike-path corridor lets you walk or cycle from the edges of town to the park's central trailhead, which is otherwise accessible by car from Highway 550. Start by crossing the historic railroad bridge. From here, the bike path parallels the beautiful Uncompahgre River, where it's not uncommon to catch glimpses of wildlife ranging from deer and foxes to bald eagles and blue herons. Once you reach the Dennis Weaver Memorial Park trailhead, follow the turn-by-turn directions to hike up through fields of fragrant sagebrush toward the summit of Boot Hill. There, Dennis Weaver's son, Rick, constructed a humorous "boot hill cemetery" to pay homage to his father and several characters on *Gunsmoke*. At the summit, take time to scope out the various cowboy-boot faux tombstones and Gunsmoke-themed epitaphs, as well as the awe-inspiring views of the Cimarron mountain range to the east and the Sneffels range (including its prominent and eponymous crown jewel, the 14,157-foot Mount Sneffels) to the south.

After you complete the loop in the park and reach the Boot Hill trailhead again, take time to relax for a while on the banks of the river. There's a 2,800-pound bronze eagle sculpture to admire and an astronomically aligned medicine wheel surrounding it, decorated with cairns, prayer stones, and poetry. From here, it's not far to stroll back to town, where bar stools, picnic tables, and a host of friendly locals at the Colorado Boy brewpub will help you feel right at home.

TURN BY TURN DIRECTIONS:

1. Head north on the bike path to cross the bridge over the river.
2. Follow the bike path for 0.8 miles, then go left over the bridge and make a right onto the Dennis Weaver Park Stagecoach Trail.
3. At 1.06, go left at the signed junction, following signs for Boot Hill Summit.
4. At 1.14, go right.
5. At 1.5, reach Boot Hill summit. Continue down over the backside onto the River Sage Scenic Loop Trail. Stay right to stay on the singletrack.
6. At 2.5, stay straight, then make a right at 2.6 to finish the lollipop loop.

FIND THE TRAILHEAD

This hike incorporates a walk on a bike path from the edge of town to the official Boot Hill trailhead in Dennis Weaver Memorial Park, so start by making your way to the north end of Railroad Street. There is a small gravel parking lot next to the bike path you'll take.

COLORADO BOY PUB AND BREWERY

Pull up a stool at this "wee humble pub," located in a historic 1915 brick building in the old section of downtown Ridgway. The seating area indoors is small, but seats at the oak bar provide glimpses of the brewing equipment. (On nice days, picnic tables let crowds spill out onto the town sidewalks.) The brewpub relies on wind-powered electricity and hot water from solar collectors on the roof to make its craft ales and artisan pizza. Growlers and 32-oz. "cowboy cans" are available to go.

CONTACT INFORMATION
Town of Ridgway,
201 N. Railroad St.,
Ridgway, CO 81432;
970-626-5738

BREWERY/RESTAURANT
Colorado Boy Pub and Brewery
602 Clinton Street,
Ridgway, CO 81432
970-626-5333
Miles from trailhead: 0.5

MONTROSE

BLACK CANYON NATIONAL PARK'S UNFATHOMABLY STEEP CLIFFS

▷··· STARTING POINT	···✕ DESTINATION
SOUTH RIM VISITOR CENTER	**OAK FLAT**
🍺 BEER	🎲 HIKE TYPE
JAZZY RAZZY	**MODERATE-STRENUOUS**
$ FEES	SEASON
$15 OR NATIONAL PARKS PASS; FREE IN WINTER	YEAR-ROUND (SNOWSHOES IN WINTER)
⛰ MAP REFERENCE	🐾 DOG FRIENDLY
BLACK CANYON PARK MAP AT NPS.GOV/BLCA	**NO**
🕐 DURATION	↦ LENGTH
1 HOUR	**1.3 MILES**
↕ LOW POINT / HIGH POINT	〰 ELEVATION GAIN
7,835 FEET / 8,186 FEET	**311 FEET**

 ALE

 CLEAR PINK

 WHEAT, RASPBERRIES

 RASPBERRIES; ZESTY

BITTERNESS **SWEETNESS**
IBU: 21

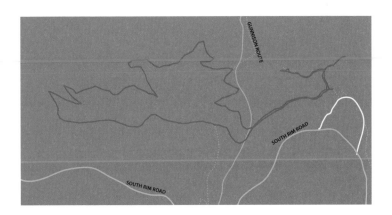

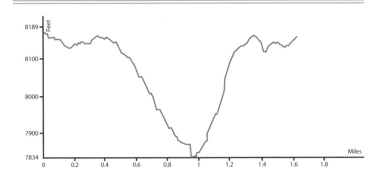

DESCRIPTION OF THE ROUTE

Home to Colorado's tallest sheer cliffs, this unfathomably steep canyon of Precambrian rock provides lovely explorations on foot any time of year.

To get a good sense of what the Black Canyon of the Gunnison is like, imagine that the Grand Canyon was narrower, steeper, and largely devoid of crowds. Carved out of otherwise unassuming rural hinterlands in western Colorado, the canyon is a vertical wilderness of Precambrian gneiss and schist more than two million years in the making. It earned its name from the fact that parts of the gorge receive only half an hour of sunlight per day; its depths are often shrouded in shadows.

In 1999, the canyon's status changed from National Monument to National Park. Due to its exceptionally steep walls, there are no marked or maintained hiking routes into the inner canyon. (Routes do exist, but all require Class 3 climbing or scrambling, route finding, and tricky footing on ladders or chained sections.) For a milder day hike, the Oak Flat Trail on the South Rim still lets you dip into the upper reaches of the canyon. Still, this lollipop loop from the South Rim Visitor Center does have some narrow sections with steep drop-offs, so only attempt it if you're comfortable with heights.

A lovely hike for much of the year, it can also be done with snowshoes in the winter—or, depending on recent snowfall, other traction devices such as Yaktrax; check ahead with the visitor center to see if snowshoe rentals are available.

As you wander down through the junipers, sagebrush, and scrub oak, you'll pass many lookout points that provide enchanting glimpses into the inner canyon. At times, you'll also be able to hear the roar of the Gunnison River echoing off the canyon walls below. By going clockwise around the loop, you'll start with some switchbacks down through the woods and finish with a steep climb back up. After you've completed your loop, don't miss the awe-inspiring views at Gunnison Point just behind the visitor center.

If you'd like to tack on another mile or two of hiking along the canyon's edge, pick up the Rim Rock Trail on the east side of the visitor center parking lot. (Note that other than in winter, leashed dogs are permitted on the Rim Rock Trail—so, if you happen to have Fido with you, this can also be a good alternative to the Oak Flat Loop where dogs are prohibited.) This trail largely parallels the road into the visitor center and offers on-foot access to a smattering of impressive overlooks of the canyon, including the majestic Tomichi Point.

TURN BY TURN DIRECTIONS:

1. Start down the wooden steps off the back deck of the visitor center.
2. Stay left, following signs for Oak Flat Loop Trail. Go left for a clockwise loop.
3. At 0.2 miles, stay straight.
4. At 1.05, go right (up), following Oak Flat Loop sign.
5. At 1.2, rejoin original trail by going left at sign pointing back to the visitor center.
6. At the visitor center, go left to check out two additional viewpoints on the canyon's edge.

FIND THE TRAILHEAD

Drive east out of downtown Montrose on Main Street/Highway 50. Go roughly 7 miles before making a left on Highway 347, following signs for Black Canyon of the Gunnison National Park. Arrive at South Rim Visitor Center in 5.4 miles.

HORSEFLY BREWING COMPANY

Horsefly Brewing is Montrose's original brewery and restaurant. It's perfectly situated on the east edge of town as you return from adventuring in the Black Canyon of the Gunnison. This refreshing fruit beer is just the ticket after a hot day in the canyon, and certainly lives up to their "No Crap on Tap" motto. A cozy interior and outdoor patio offer plenty of welcoming space to enjoy your ales and some traditional brewpub grub.

CONTACT INFORMATION
South Rim Visitor Center,
9800 Highway 347,
Montrose, CO 81401;
970-641-2337

BREWERY/RESTAURANT
Horsefly Brewing Company
846 E Main St,
Montrose, CO 81401
970-249-6889
Miles from trailhead: 13

TELLURIDE

HIKE ALONGSIDE THE TUMBLING RAPIDS OF THE SAN MIGUEL RIVER

▷⋯ STARTING POINT	⋯✕ DESTINATION
TRAILHEAD ON SAN MIGUEL RIVER ROAD	**KEYSTONE GORGE**
🍺 BEER	HIKE TYPE
FACE DOWN BROWN ALE	**MODERATE**
$ FEES	SEASON
NONE	**APRIL TO OCTOBER**
⛰ MAP REFERENCE	🐾 DOG FRIENDLY
KEYSTONE GORGE LOOP TRAIL ON SANMIGUELCOUNTYCO.GOV	**NO**
🕐 DURATION	↦ LENGTH
1-1.5 HOURS	**2.5 MILES**
↑↓ LOW POINT / HIGH POINT	〰 ELEVATION GAIN
8,182 FEET / 8,710 FEET	**528 FEET**

 BROWN ALE

 DARK BROWN

MALTY; CHESTNUTS, TOFFEE

SMOOTH; ROASTED COFFEE BEANS

BITTERNESS
IBU: 37

SWEETNESS

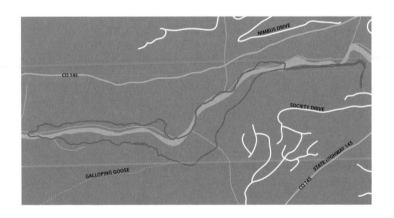

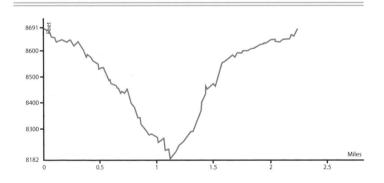

DESCRIPTION OF THE ROUTE

Hike along the tumbling rapids of the San Miguel River before sampling an award-winning brown ale made from the same snow melt that fuels the river.

The picturesque mountain town of Telluride sits tucked away into the back of a jaw-dropping box canyon. It's known as much for its epic ski runs as its summer festivals, including favorites like the Telluride Bluegrass Festival and, for the beer aficionados, Telluride Blues & Brews. Beer culture comes baked in to most ski towns anyway, but not every ski town boasts a brewery that regularly earns the kind of international accolades that Telluride Brewing Company reels in.

One of the highest microbreweries in the world at 8,750 feet, Telluride Brewing Company sits three miles west of downtown Telluride in a satellite community called Lawson Hill. It's on the west end of the Valley Floor, a large tract of land acquired in 2007 by the town of Telluride to be preserved for the public and undeveloped.

While tourists often flock to the iconic day hikes accessible from town itself—Bear Creek and the Jud Wiebe—the Keystone Gorge loop offers a quiet respite from such crowds. This lovely lollipop loop dips down into a river gorge, rather than climbing in elevation, which makes it a great early- or late-season hike when other trails are snowed in. It crams in a lot in a few short miles, including relics from the town's mining days—scattered artifacts and ruins, as well as several abandoned mine entrances.

Shortly after crossing the first footbridge over the river, you'll catch a fleeting glimpse of Wilson Peak, which should look familiar if you've ever looked at a can of Coors beer. As the singletrack winds down along the river, you'll toe the edges of high bluffs before dropping down next to the river. Watch your footing, as the trail grows narrow and loose in places. The trail hugs the river's edges, so enjoy the clamor of the rapids by your side as they gush over boulders and spill toward the neighboring Ilium Valley. The woods here are lush and cool.

At the lower suspension bridge, there's a picnic table and some rock beaches for relaxing in the glow of the afternoon sun. On the south side of the loop, you'll pass through more conifer forest before popping out into stands of aspen that look aflame come late September or early October. Pop out on the railroad-grade Galloping Goose Trail (named for the motorized railcars that once ran here, when the Rio Grande Southern Railroad still existed in the 1950s), which will take you back to your starting point where the brewery awaits.

TURN BY TURN DIRECTIONS:

1. Go down the gravel hill and stay left to parallel the river on doubletrack going west.
2. At 0.25 miles, go right over the bridge, then left to follow Keystone Gorge loop.
3. At 1.1, cross the lower bridge and hang a left.
4. At 1.6, go left onto the Galloping Goose Trail.
5. At 2, pass the original junction with the upper bridge; stay straight to finish the lollipop loop.

FIND THE TRAILHEAD

From the town of Telluride, drive west for 3 miles, make a left at the roundabout onto Highway 145, then a right onto Society Drive. After passing the brewery, turn right onto San Miguel River Road. The trailhead, signed for the Galloping Goose and Keystone Gorge, is just behind the Lawson Hill bus circle on your right-hand side. It's also possible to take the free Galloping Goose bus here, or walk or bike from town via the paved bike path or singletrack trails across the Valley Floor.

TELLURIDE BREWING COMPANY

Formed by best friends, adventure buddies, and longtime beer enthusiasts Chris Fish and Tommy Thacher, Telluride Brewing Company opened its doors in 2012. Its popular Face Down Brown has taken home the gold three times at GABF (Great American Beer Festival) and the World Beer Cup. Though TBC has greatly expanded its production and distribution all across Colorado, they have no plans to distribute out of state (at least as of the writing of this book).

CONTACT INFORMATION
San Miguel County Parks & Open Space,
333 W Colorado Avenue,
3rd Floor, Telluride, CO 81435;
970-369-5469

BREWERY/RESTAURANT
Telluride Brewing Company
156 DEF Society Drive,
Telluride, CO 81435
970-728-5094
Miles from trailhead: 0.2

GRAND JUNCTION

A TASTE OF THE DESERT THROUGH COLORFUL CANYONS

▷⋯ STARTING POINT	⋯✕ DESTINATION
LUNCH LOOP TRAILHEAD ON MONUMENT ROAD	**LUNCH LOOPS**
🍺 BEER	🎛 HIKE TYPE
BLACK'S BRIDGE STOUT	**MODERATE**
💲 FEES	📅 SEASON
NONE	**FEBRUARY TO NOVEMBER**
🗺 MAP REFERENCE	🐾 DOG FRIENDLY
LUNCH LOOP TRAILS AT COPMOBA.ORG	**YES**
🕐 DURATION	↦ LENGTH
2-2.5 HOURS	**5 MILES**
↑↓ LOW POINT / HIGH POINT	〰 ELEVATION GAIN
4,672 FEET / 5,394 FEET	**690 FEET**

STOUT

BLACK;
LARGE FOAMY HEAD

MOCHA; EARTHY

ROASTED COFFEE;
CREAMY CHOCOLATE

BITTERNESS   SWEETNESS
IBU: 40

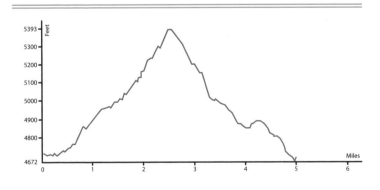

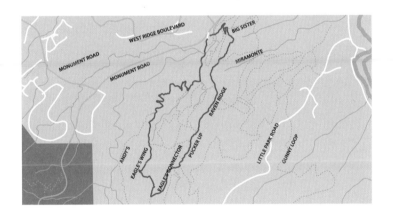

DESCRIPTION OF THE ROUTE

Get a taste of the desert with this roller coaster of a loop up, down, over, and through the most colorful of canyons.

For mountain bikers, hikers, and trail runners who reside in the high mountains of western Colorado, it's an early spring tradition to head west to "the desert" of Grand Junction. Because of its relatively low elevation and desert-like climate, GJ provides plenty of bare singletrack to play on long before the trails in the high country have melted out. Located on Bureau of Land Management (BLM) land, the Tabeguache trails—or, as they're commonly called, the Lunch Loops—link up dozens of miles of singletrack geared primarily at mountain bikers. However, they make for lovely day hiking as well. When you're done, Kannah Creek's Edgewater Brewery sits just across the river.

This hike takes you through beautiful, colorful canyons filled with boulders, bentonite clay, broken rock slabs, and desert grasses. Vast views of the surrounding canyons grace your line of sight for much of the way. Trees are few and far between, and shade is non-existent; wear sunscreen and bring plenty of water.

The trails are pretty well marked along the way, including all major intersections—so, despite the complicated turn by turn directions, it would be hard to get lost. Nonetheless, it's useful to print out a trail map from the Colorado Plateau Mountain Bike Trail Association (COPMOBA) website, or pick up one in a bike or outdoor shop in town.

This particular loop links up several of the system's popular trails, including the Eagle's Tail, Wing, and Connector trails, Tabeguache, Lemon Squeezer, Curt's Lane, and Curt's Down. You'll understand why it's called Lemon Squeezer when the trail seems to disappear entirely. "Squeeze" yourself between the two large boulders to continue downhill, finding your way on the other side.

There are plenty of other trails beyond the ones in this loop to explore, but do be careful not to venture onto bike-only trails, or go the wrong direction on one-way routes. The Lunch Loops are best hiked in the shoulder seasons and, ideally, not after any heavy rains, which can wreak havoc on the trails. If you're hiking in the summer, start early in the day, as this fully exposed trail system can feel like a furnace by high noon. Remember to be mindful of other trail users, and enjoy some of western Colorado's finest desert hiking!

TURN BY TURN DIRECTIONS:

1. Start on wide gravel path to the right of the bike park.
2. At 0.4 miles, go right onto Eagle's Tail, then right at 1.85 onto Eagle's Wing and left at 2.5 onto Eagle's Connector.
3. At 2.7, go left on Tabegauche, then right at 3.1, following signs for Lemon Squeezer.
4. At 3.4, go straight onto Curt's Lane, then straight onto Curt's Down at 4.2.
5. At 4.5, go right, then stay left to return to the trailhead (within sight).

FIND THE TRAILHEAD

Take Broadway (Highway 340) west out of Grand Junction. Shortly after crossing over the Colorado River, make a left onto Monument Road. In 1.5 miles, arrive at the Lunch Loop Trailhead on your left.

KANNAH CREEK BREWING COMPANY

Kannah Creek boasts several different locations in western Colorado, including its Edgewater brewpub situated on a spacious lawn on the banks of the Colorado River. Guinness lovers will rejoice at Kannah Creek's nitrogenized stout—a rich, foamy, smooth-as-butter beer that's not overly sweet. Come for the beer; stay for the pub fare, live music, and corn hole.

CONTACT INFORMATION
BLM,
Grand Junction Field Office,
2815 H Road,
Grand Junction, CO 81506;
970-244-3000

BREWERY/RESTAURANT
Kannah Creeek Brewing Company
905 Struthers Avenue,
Grand Junction, CO 81501
970-263-0111
Miles from trailhead: 3.9

PALISADE

A CLIFF-EDGE TOUR OF ANCIENT PETROGLYPHS AND ROCK FORMATIONS

▷⋯ STARTING POINT	⋯✗ DESTINATION
RAPID CREEK TRAILHEAD	**PALISADE RIM**
🍺 BEER	HIKE TYPE
DIRTY HIPPIE DARK AMERICAN WHEAT	**STRENUOUS**
$ FEES	SEASON
NONE	**FEBRUARY TO NOVEMBER**
🏔 MAP REFERENCE	🐾 DOG FRIENDLY
PALISADE RIM TRAIL ON COPMOBA.ORG	**YES**
🕐 DURATION	↦ LENGTH
2-3 HOURS	**5.2 MILES**
↑↓ LOW POINT / HIGH POINT	〰 ELEVATION GAIN
4,724 FEET / 5,617 FEET	**850 FEET**

AMERICAN WHEAT

 MEDIUM-DARK BROWN

 FAINT CARAMEL

LIGHT, MALTY, CHOCOLATE

BITTERNESS	SWEETNESS
IBU: 17	

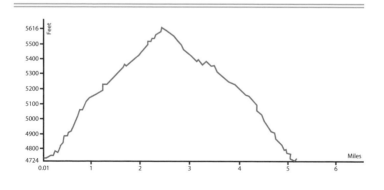

DESCRIPTION OF THE ROUTE

Hike high above the vineyards and orchards of Grand Valley on a cliff-edge tour of ancient petroglyphs and rock formations.

There may be no better part of Colorado to work up your appetite and thirst than the town of Palisade. Thanks to its relatively low elevation, sunny climate, and six-month-long growing season, it's well situated for the many farms, ranches, vineyards, orchards and roadside fruit stands that call Palisade home. From July to September, local orchards harvest countless tons of giant, juicy Palisade peaches that are famous and highly sought all over the American West.

This cliff-side jaunt offers the best bird's-eye views of Palisade's patchwork farms and vineyards and the Colorado River below. Utilizing trails built only recently (read: not yet on many people's radar screens) after the granting of a land easement for public recreational use, this loop provides near-constant panoramic views of the Grand Valley below. It also passes by a series of ancient petroglyphs carved into rock faces.

The hike begins with a brief, flat section paralleling the river. Duck under the giant cottonwood tree next to the trailhead sign to find your route, and enjoy the easy terrain while it lasts before crossing the highway on your left. Pick up the signed singletrack that begins winding up the canyon. From here, you'll have a steep climb up through gnarled, twisty junipers into an impressive amphitheater of high rock canyons and bluffs. About three quarters of a mile in, keep an eye out for the improbably balanced boulder in the gully below you. Once you reach the higher plateau, the trail mellows to an easier grade while winding along cliff edges. Some areas of extreme exposure do exist, so do not attempt if you have a fear of heights. Otherwise, it's hard to overstate the magnificence of the views from the steep slopes the trail traverses.

Around mile 1.5, look for small social trails that detour off the main route over to the petroglyphs—well worth a side trip to check out.

Several options exist: you can complete just the Lower Loop Trail for a 4-mile lollipop loop, the Upper Loop Trail for a 9.3-mile double lollipop loop (including longer sections of exposed, cliff-edge hiking), or the hybrid compromise outlined here that includes the Lower Loop plus an out-and-back on the connector trail.

As with many hikes in the desert environs of western Colorado, this trail was largely developed by and for mountain bikers. Though hiking and equestrian use is permitted, do take extra care to remain aware of your surroundings and be respectful of other trail users. Wear a hat and sunscreen, carry ample water, and avoid this trail after heavy rains or during the height of the day in extreme summer heat.

TURN BY TURN DIRECTIONS:

1. Start hiking along the river from the southeast corner of the parking area.
2. At 0.1 miles, cross the highway on your left and pick up the signed singletrack.
3. At 1.1, go right at the junction to initiate a counterclockwise loop.
4. At 1.7, go right, following signs for Upper Rim Trail.
5. At 2.45, you have an option to extend your hike by 4 miles by doing the upper loop. Otherwise, turn around and head back.
6. At mile 4 (or 8 if you've done the upper loop), go right to complete your lollipop loop.

FIND THE TRAILHEAD

From Palisade, drive northeast out of town on North River Road. In about 2 miles, make a right onto Highway 6, then make an immediate right into the Rapid Creek parking area on the left, next to the visitor information structure by the river.

PALISADE BREWING COMPANY

Palisade is a place primarily known for its fruit (peaches especially) and wine. But, as Palisade Brewing Company's owners love to ask, "Who says you can't make beer in wine country?" With a focus on American-style ales and barbecue, this brewery and its sun-drenched outdoor beer garden are a perfect place to relax, refuel, and enjoy views of the surrounding foothills and rock formations. Enjoy this sessionable Dirty Hippie Dark American Wheat with a slice of orange.

CONTACT INFORMATION
BLM, Grand Junction Field Office,
2815 H Road,
Grand Junction, CO 81506; 970-244-3000

BREWERY/RESTAURANT
Palisade Brewing Company
200 Peach Ave,
Palisade, CO 81526
970-464-1462
Miles from trailhead: 2.7

GLENWOOD SPRINGS

A CLASSIC, HISTORIC LOOP HIKE FROM DOWNTOWN

▷⋯ STARTING POINT	⋯✗ DESTINATION
SCOUT TRAILHEAD	**LOOKOUT MOUNTAIN; DOC HOLLIDAY'S GRAVE**
🍺 BEER	🎲 HIKE TYPE
DOS RIOS VIENNA LAGER	**STRENUOUS**
$ FEES	📅 SEASON
NO FEES	**MARCH TO NOVEMBER**
⛰ MAP REFERENCE	🐾 DOG FRIENDLY
TRAILS ILLUSTRATED 151: FLAT TOPS SOUTH	**YES (LEASH REQUIRED)**
🕐 DURATION	↦ LENGTH
4-5 HOURS	**7.5 MILES**
↑↓ LOW POINT / HIGH POINT	∿ ELEVATION GAIN
5,912 FEET / 8,074 FEET	**2,090 FEET**

 LAGER

 CLEAR RED

GRAINS, FAINT MALT

SMOOTH AND TANGY

BITTERNESS
IBU: 25

SWEETNESS

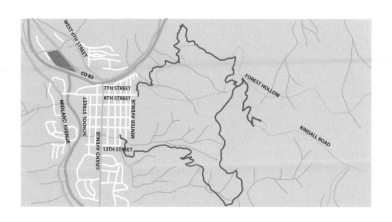

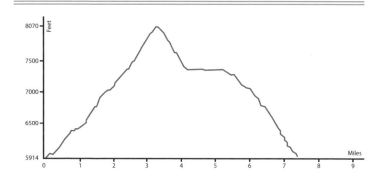

DESCRIPTION OF THE ROUTE

This loop hike is well loved by locals for its easy accessibility from downtown, beautiful views and historic landmarks.

The Scout Trail—or "Boy Scout Trail," as it's called by locals—is a classic Glenwood hike that sits just a few blocks away from the Glenwood Canyon Brewpub, one of Colorado's original breweries. The Scout Trail is a historic path that the Ute Indians once used to access the area's natural hot springs. The springs' 124-degree mineral waters are what first put Glenwood on the map and now also supply the town's popular, often raucous (albeit family-friendly) Hot Springs Pool and Spa.

As an out-and-back, the Scout Trail is just under 5 miles (or 6.5 if you go all the way to Lookout Mountain Park). But, for a few extra smooth, relatively easy miles, it's possible to turn this hike into a loop that concludes near another iconic Glenwood landmark: the grave of infamous gambler and gunfighter "Doc" Holliday.

Holliday traveled to Glenwood in his final days, hoping (in vain, ultimately) that the hot springs could help treat his tuberculosis. The Pioneer Cemetery trailhead near his tombstone is just a few blocks away from the Scout trailhead where you started, as well as the Glenwood Canyon Brewpub, where you'll finish.

After a steep initial climb through pinyon and juniper, you'll emerge in a high meadow of sagebrush and scrub oak. Along the way, enjoy great views of the iconic Horseshoe Bend in the Colorado River and the steep cliffs of Glenwood Canyon. Don't be alarmed if you hear screams either; they likely belong to the hearty-souled riders of the Giant Canyon Swing at Glenwood Caverns Adventure Park, perched high atop the bluffs across the valley, and visible from much of the Scout Trail.

From the top of Lookout Mountain Park, scope out local mountains like Capitol Peak, Mount Sopris, Chair Mountain, and Glenwood's own Sunlight Ski Area. Don't neglect to pop your head inside the historic outhouse just below the radio tower. More than 80 years old and no longer functional, but preserved by the BLM as a cultural resource, it's certain to be one of the more spectacularly positioned commodes you've ever come across!

Many splintering trails exist on the patch of BLM land that these trails use, so it is wise to carry a map. Also, stay alert for mountain bikers.

TURN BY TURN DIRECTIONS:

1. At mile 2.4, go left at the signed intersection, then make a right shortly thereafter (not straight onto Forest Hollow) to climb up to Lookout Mountain.
2. At 3.1, reach a fire ring and unmarked trail intersections. Go straight through to visit the historic outhouse, or hang a left to climb up to the radio towers and the best views.
3. Return down the way you came until you again reach the junction with Forest Hollow. Go left (also the way you came) here, then either right (the way you came) for an out-and-back, or straight (new trail) to initiate a longer loop over to Doc Holliday's grave.
4. If taking the longer loop, follow the main route by going right at 5.4, right (through a fence opening) at 5.6, right at 6.1, right again at 6.2, then left and steep down at 6.8 until you reach the cemetery, four blocks south of the Scout trailhead.

FIND THE TRAILHEAD

From downtown Glenwood Springs, travel east on 8th Street for several blocks until it dead-ends at the Scout trailhead.

GLENWOOD CANYON BREWPUB

Founded in 1996, Glenwood Canyon Brewpub was built in the historic Hotel Denver, located just across the Colorado River from Glenwood's infamous hot springs pool. It boasts tasty pub fare (try the beer-battered onion rings or beer cheese soup made with their Vapor Cave IPA) and numerous award-winning beers—many named after local trails, mountains, river rapids or other locals' favorite outdoor haunts.

CONTACT INFORMATION
BLM,
Colorado River Valley Field Office,
2300 River Frontage Road,
Silt, CO 81652;
970-876-9000

BREWERY/RESTAURANT
Glenwood Canyon Brewpub
402 7th Street,
Glenwood Springs, CO 81601
970-945-1276
Miles from trailhead: 0.3

CARBONDALE

GET UP HIGH ABOVE TOWN ON RED HILL'S MOST PROMINENT LOOKOUT

▷⋯ STARTING POINT	⋯✕ DESTINATION
RED HILL TRAILHEAD	**MUSHROOM ROCK**
🍺 BEER	HIKE TYPE
FREESTONE EXTRA PALE ALE	**STRENUOUS**
$ FEES	SEASON
NONE	**MARCH TO NOVEMBER**
⛰ MAP REFERENCE	🐾 DOG FRIENDLY
RED HILL TRAIL MAP AT ASPENTRAILFINDER.COM	**YES**
🕐 DURATION	↦ LENGTH
1-1.5 HOURS	**2 MILES**
↑↓ LOW POINT / HIGH POINT	〰 ELEVATION GAIN
6,138 FEET / 6,978 FEET	**875 FEET**

PALE ALE

 BRIGHT GOLD

 PINE, EARTHY HOPS

 DRY, CRISP; CITRUS

BITTERNESS IBU: 38	SWEETNESS

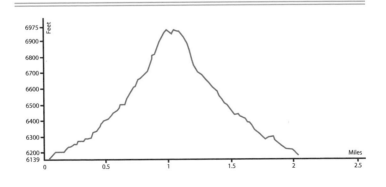

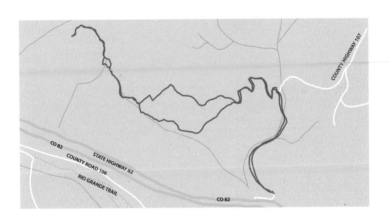

DESCRIPTION OF THE ROUTE

Get up high above town on this lung-searing gut buster of a hike and enjoy a front-row seat to admire one of Colorado's most prominent peaks.

The aptly named Red Hill Special Recreation Management Area sits north of downtown Carbondale, a quaint, artsy mountain town nestled in ranching country at the foot of Colorado's Elk Mountains. The area's red rocks, sagebrush meadows, and juniper trees are home to some 19 miles of singletrack, as well as herds of deer and elk.

From the trails that wend steeply up Red Hill's southern-facing walls, you'll enjoy many views—the town of Carbondale itself, the convergence of the mighty Roaring Fork and Crystal rivers, and poised in stunning prominence to the south, the frequently snow-capped Mount Sopris. With its twin summits clocking in at 12,965 feet, Sopris does not reach as high as many of its more famous neighboring 13er and 14er peaks, but its near 7,000 feet of relative prominence (the total elevation change from a peak's base to its summit) make it one of the largest mountains in the contiguous United States. And there's no better place from which to ogle the looming giant than Mushroom Rock, a beautiful red-rock formation you can clamber your way steeply up to on the southern flanks of Red Hill.

The trail gets a bit splintered in places toward the top, but do your best to prevent further soil erosion by staying on the main route. It's a calf burner, to be sure, but well worth it for the views you'll enjoy along the way, not to mention the sense of accomplishment you'll feel splayed out atop the fiery red Mushroom Rock.

If you'd like to extend your hike, many options exist up top, once you've gotten the initial climb over with; carry a trail map, or snap a photo of the one at the trailhead to use for reference. Tacking on Skeeters Ridge and a loop of the Faerie and Bogus trails, followed by a descent back to the trailhead on Three Gulch, offers a great way to extend your hike along the cliffs and through the woods and sagebrush. Otherwise, it's a quad burner of a descent back down to the bottom, where fresh, seasonal brews at Batch by Roaring Fork Beer Company will be calling your name.

TURN BY TURN DIRECTIONS:

1. Walk up the dirt road a quarter-mile and look for the Red Hill trailhead sign on your left.
2. Stay left to follow the Three Poles Trail.
3. Stay left at the junctions at 0.5 and 0.75 miles in, following signs for Mushroom Rock.
4. At 1 mile, reach Mushroom Rock. Return the way you came, or for a slight change of scenery, go left at 1.3 miles, then right at 1.4, before returning on the trail you came up.

FIND THE TRAILHEAD

From downtown Carbondale, go north on Highway 133. The Red Hill trailhead can be found on the north side of the intersection of Highway 82 and Highway 133. Park in the large dirt lot.

ROARING FORK BEER COMPANY

With a focus on seasonal, experimental small-batch beers, RFBC quickly outgrew its original space adjacent to Red Hill, so it has since expanded into a new downtown location. The Batch tasting room sports 12 in-house taps with everything from barrel-aged brews to imperial IPAs and grapefruit- or watermelon-infused renditions of the Freestone Extra Pale Ale. Additionally, it features guest taps for other select Colorado craft beers, wines (including a dry-hopped sauvignon blanc by Denver winery Infinite Monkey Theorem), hard cider, kombucha, craft sodas and even sake.

CONTACT INFORMATION
BLM,
Colorado River Valley Field Office,
2300 River Frontage Road,
Silt, CO 81652;
970-876-9000

BREWERY/RESTAURANT
Roaring Fork Beer Company
358 Main Street
Carbondale, CO 81623
970-510-5934
Miles from trailhead: 1.8

ASPEN

WOODS, ABANDONED MINE SHAFTS, AND EPIC VIEWS

▷··· STARTING POINT	···✕ DESTINATION
HUNTER CREEK TRAILHEAD	**SMUGGLER MOUNTAIN**
🍺 BEER	🔲 HIKE TYPE
SILVER CITY KETTLE-SOURED SESSION ALE	**MODERATE**
$ FEES	📅 SEASON
NONE	**APRIL TO OCTOBER**
⌖ MAP REFERENCE	🐾 DOG FRIENDLY
WWW.PITKINOUTSIDE.ORG	**YES (LEASH REQUIRED)**
🕐 DURATION	↦ LENGTH
2-3 HOURS	**4.1 MILES**
↑↓ LOW POINT / HIGH POINT	〽 ELEVATION GAIN
7,933 FEET / 9,029 FEET	**1,050 FEET**

 SESSION ALE

 CLEAR, LIGHT GOLD

 FLORAL, APPLE, PINE

 LIGHT, TART, LEMON, WHEAT

BITTERNESS	SWEETNESS
IBU: 14	

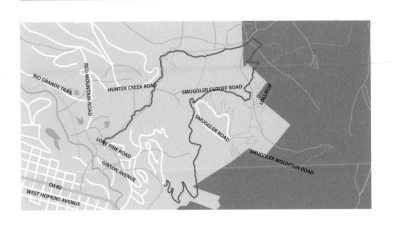

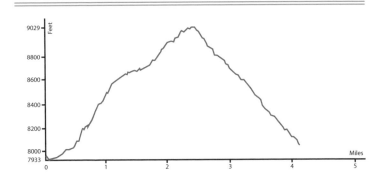

DESCRIPTION OF THE ROUTE

Take a "hiking tour de Aspen" through the woods, past an abandoned mine shaft, up to the notorious Smuggler Mountain observation deck, and back down to town for a kettle-soured session ale that honors Aspen's silver-mining heritage.

It's easy to get overwhelmed by the multitude of trails that radiate out of Aspen like a massive spider web. Before venturing too far into the beckoning high-altitude backcountry, get acclimated and oriented with this popular hike right out of town.

This loop combines hiking up the Hunter Creek and Hunter Creek Cutoff trails with a descent down the popular Smuggler Mountain Road, which overlooks all of Aspen, the ski area, and many of the surrounding peaks.

You'll begin on a very well-shaded trail through lush, mixed forest and along Hunter Creek. Watch your footing over the rocks and roots on the fun, twisting singletrack. At times, you'll also be on wooden boardwalks, rock staircases, or crossing bridges back and forth over the creek. After venturing onto the quieter Cutoff trail, enjoy passing through open meadows and dazzling aspen trees. Watch out for mountain bikers here.

After turning off toward Smuggler Mountain, you'll pass the long-abandoned Iowa Shaft Mine, complete with informational signs and a map of the historical site. Shortly after, check out the Smuggler observation deck—a lovely wooden platform that proffers a terrific view of tiny Aspen below. From here, it's an easy walk down the dirt Smuggler Mountain Road back to town. Enjoy the people-watching-dog walkers, mountain bikers, yoga-pants-clad locals, winded tourists, fellow hikers, and more.

When you get down to the Smuggler Mountain trailhead, it's just half a mile back to the Hunter Creek trailhead (if you drove and left a car there), or less than a mile to meander into town and grab yourself a barstool at the Aspen Brewing Company taproom. There, an easy-drinking session ale named in honor of the city's silver-mining heritage awaits you.

TURN BY TURN DIRECTIONS:

1. Cross five bridges in the first mile.
2. At mile 0.85, go right to cross the fifth bridge (Benedict Bridge) and stay on the main trail up.
3. At 1.5, turn right onto the well-signed Hunter Creek Cutoff trail. Keep following signs for Hunter Creek Cutoff.
4. At 2.2, make a right, following signs for the observation deck and Smuggler Mountain Road.
5. At 2.6, go straight across the trail intersection to check out the observation deck before returning to the intersection and walking 1.5 miles down the wide, dirt Smuggler Road.

FIND THE TRAILHEAD

No car necessary, though there is a small parking area at the Hunter Creek trailhead. Either take the free Hunter Creek shuttle or walk north out of town on Mill St (which turns into Red Mountain Rd) and make a right on Lone Pine Rd; the well-marked Hunter Creek trailhead will be on your left shortly.

ASPEN BREWING COMPANY

Aspen Brewing Company loves to boast that its beer is made "downstream from nobody"—and, perched right around 8,000 feet above sea level and surrounded on all sides by mountains, they're not wrong. Its taproom is a small, boisterous space located in the heart of downtown Aspen—just a few blocks away from both the starting and finishing trailheads of your hiking tour. This unique kettle-soured and dry-hopped session ale is "low in alcohol and big in flavor"—just what you need to not knock yourself out imbibing at altitude!

CONTACT INFORMATION
U.S. Forest Service,
Aspen-Sopris Ranger District,
806 W Hallam Street,
Aspen, CO 81611;
970-963-2266

BREWERY/RESTAURANT
Aspen Tap/Aspen Brewing Company
121 S. Galena Street
Aspen, CO 81611
970-710-2461
Miles from trailhead: 0.6

CRESTED BUTTE

A SECLUDED BASIN WITH WILDLIFE, WATERFALLS, AND WILDFLOWERS

▷··· STARTING POINT	···✕ DESTINATION
COPPER CREEK TRAILHEAD	**JUDD FALLS, COPPER LAKE, OR EAST MAROON PASS**
🍺 BEER	HIKE TYPE
GERMAN PILSNER	**MODERATE-STRENUOUS**
$ FEES	SEASON
NONE	**JULY TO OCTOBER**
MAP REFERENCE	🐾 DOG FRIENDLY
TRAILS ILLUSTRATED 131: CRESTED BUTTE, PEARL PASS	**YES (LEASH REQUIRED)**
⏲ DURATION	↦ LENGTH
5-8 HOURS	**10.5-12.4 MILES**
↑↓ LOW POINT / HIGH POINT	〰 ELEVATION GAIN
9,603 FEET / 11,437-11,864 FEET	**2,100-2,410 FEET**

PILSNER

LIGHT, GOLD;
UNFILTERED

FLORAL, SPICY,
HERBAL

CRISP; TOASTED
CRACKER

BITTERNESS   SWEETNESS
IBU: 33

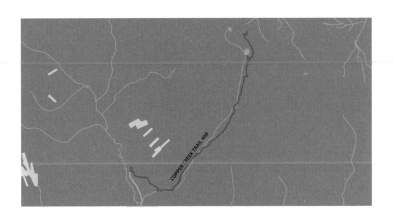

COPPER CREEK TRAIL 468

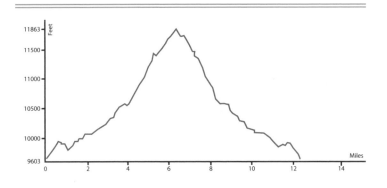

DESCRIPTION OF THE ROUTE

This secluded basin has it all, from wildlife to waterfalls to wildflowers—and when you're done, grab a German-style pilsner back in town.

On a dirt road just north of Crested Butte lies the tiny mountain hamlet of Gothic—an abandoned mining town turned high-altitude field station for biological scientists and researchers. Behind Gothic lies a massive basin teeming with trails and craggy peaks. The area is renowned for its incredible wildflowers in the summer (which typically peak in July into early August) and fiery-hued leaves in the autumn.

The Copper Creek Trail is a highlights reel of some of the best scenery the basin has to offer, from waterfalls to lakes to wildflowers.

Many people make the stroll through the aspen copse and across the hillside for a glimpse of Judd Falls. The waterfall overlook, complete with a bench, is clearly marked with a sign and is just a few steps off the main trail. If you're not up for a long day hike, this makes a great destination and turnaround spot for a quick, scenic jaunt.

After the falls overlook, though, the crowds thin to a trickle. Continue on the Copper Creek Trail. Some tempting social trails exist, so be diligent about staying on the main trail. The route is exceptionally clear and well marked at all intersections. You'll pass through conifer forest, open meadows, avalanche chutes, and across a number of creeks and streams. Steep, rocky cliffs, and ridges hover over either side of the valley as you make your way toward the jagged massifs at the back.

The "island" at the near edge of Copper Lake makes a great lunch spot. Enjoy gazing up at the immense cirque above and admiring the wildflowers around you. From here, you can see the trail up to East Maroon Pass cutting through the talus slope to your right at a forgiving grade. This part is optional; it tacks on an extra 500 feet of climbing and one mile (two roundtrip) to tag the top of East Maroon Pass, but this bonus stretch is well worth your time and effort if you're up for it. The views and wildflowers will blow your mind.

TURN BY TURN DIRECTIONS:

1. From the lower 2WD trailhead, walk up the road half a mile to reach the official Copper Creek trailhead (far right end of 4WD parking area).
2. At 1.1 miles, reach scenic Judd Falls overlook.
3. At 1.3, pass a large sign for Copper Creek Trail. Continue straight (left of the sign) on the main route.
4. There are sizable creek crossings at miles 2.5 and 4. Your feet will probably get wet.
5. At mile 5.25, stay left to go to Copper Lake. This is a good turnaround spot, or you can continue up to East Maroon Pass.
6. If continuing, follow trail signs to the large rock pile atop East Maroon Pass at 6.35 miles, before returning the way you came.

FIND THE TRAILHEAD

In the summer, you can take the Gothic Bus Shuttle (www.mtnexp.org) from town. Otherwise, drive 8 miles north of Crested Butte on Gothic Road (County Road 371). Park at the first obvious parking area (complete with outhouse) after Gothic, or if you have a high-clearance vehicle, hang a right and proceed another half-mile up the 4WD road to the Copper Creek trailhead. (Subtract 0.5 mile from the trail mileages in the turn-by-turn directions if you do this.)

IRWIN BREWING COMPANY

Irwin Brewing Company opened its door in January 2017 with a 15-barrel brewing system to fill a brewery void in the Gunnison Valley. They launched with a lineup of six year-round beers ranging from this light pilsner to a dark, full-bodied oatmeal stout. Named after the nearby Lake Irwin, the production brewery has kegs and growler fills available; however, you'll have to head over to the Public House (same ownership) or other assorted restaurants in town to consume a fresh pint on site.

CONTACT INFORMATION
U.S. Forest Service,
Gunnison Ranger District,
216 N. Colorado Street,
Gunnison, CO 81230;
970-641-0471

BREWERY/RESTAURANT
Irwin Brewing Company
326 Belleview Avenue,
Crested Butte, CO 81224
970-275-7578
Miles from trailhead: 8.8

GUNNISON

MILES OF HIGH-DESERT WANDERING

▷⋯ STARTING POINT	⋯✗ DESTINATION
ESCALANTE DRIVE TRAILHEAD	**RIDGELINE TRAIL**
🍺 BEER	🔀 HIKE TYPE
SOL'S ESPRESSO STOUT	**EASY-MODERATE**
$ FEES	📅 SEASON
NONE	**APRIL TO NOVEMBER**
⛰ MAP REFERENCE	🐾 DOG FRIENDLY
GUNNISONTRAILS.ORG	**NO**
🕐 DURATION	↦ LENGTH
2-3 HOURS	**5.8 MILES**
↕ LOW POINT / HIGH POINT	〰 ELEVATION GAIN
7,776 FEET / 8,510 FEET	**620 FEET**

 STOUT

 BLACK; FROTHY HEAD

 CHOCOLATE, ESPRESSO

 VELVETY; ROASTED COFFEE

BITTERNESS
IBU: 33

SWEETNESS

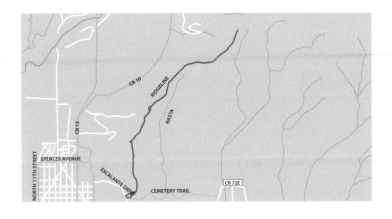

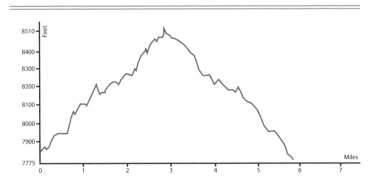

DESCRIPTION OF THE ROUTE

Largely undeveloped, this vast tract of open, high-desert BLM land is still the Wild West of recreation.

While many hikers, trail runners, and mountain bikers flock to the singletrack at Hartman Rocks several miles south of town, it's also possible to access dozens of miles of lesser-known trails within walking distance of downtown Gunnison. Just to the east of the Western State Colorado University campus sits the 9,000-foot Signal Peak, as well as expansive swaths of BLM land that surround it. The land is largely undeveloped and unmapped for recreational use, though two major routes—the Contour Trail and the Ridgeline Trail—and several smaller spur trails provide plenty of miles of high-desert wandering.

Much of this route utilizes part of the Gunnison Spur of the Colorado Trail—a lengthy extension (no longer recognized or maintained as an official part of the Colorado Trail) connecting the town of Gunnison to the Colorado Trail, a 486-mile long-distance trail that cuts across the state from Denver to Durango. On your way up toward the radio towers atop the Ridgeline Trail, you'll meander through sagebrush and desert grasses. You'll pass the rusted out, smashed up, bullet-ridden remnants of an old car that's slowly sinking into the clutches of the earth.

The views are vast in all directions, including the surrounding cliffs, nearby Tenderfoot Mountain, and up toward snow-capped West Elk Mountains. You'll recognize Tenderfoot from the iconic, white "W" on its flanks. The massive monogram was reputedly first created in the 1920s when the college, then known as the Colorado State Normal School, was officially renamed Western State. As the story goes, every May, upperclassmen incited freshmen to haul 100-pound sacks of lime up the mountain on foot to help freshly "whitewash" the letter.

Hike as far out as you wish. Going all the way to the top of Signal Peak, the high point on the horizon to your right, tacks on roughly 3.5 bonus miles and adds another 500 feet of climbing to the hike outlined here. Otherwise, the large cairn at an unmarked trail junction at mile 2.9 makes a perfectly suitable turnaround spot.

The downside to the lack of planning for recreation in this area is that it's not well mapped, and there are virtually no trail signs; junctions are unmarked. The upside is that the lack of trees makes it nearly impossible to get lost and, at least on a clear day, many visible landmarks on the horizon can help orient you. That said, the lack of trees and complete exposure of the terrain makes this area extremely dangerous during thunderstorms; avoid being on this trail at all costs if lightning is present or forecasted.

TURN BY TURN DIRECTIONS:

1. Find your route by following the Spur signs and wooden posts.
2. At 0.4 miles, veer right onto narrow singletrack to follow the Contour Trail.
3. At 0.6, stay left at the unmarked fork (following wooden posts) to get on the Old Car Trail.
4. At 0.75, pass the old rusted out car and go right onto the gravel doubletrack. This is the Ridgeline Trail.
5. At 2.9, reach a large stack of rocks at an unmarked trail junction. Turn around and return the way you came.

FIND THE TRAILHEAD

From downtown Gunnison, drive east on Tomichi Avenue/Highway 50. Turn left onto Adams Street, then right onto Georgia Avenue, which becomes Escalante Drive. In 0.3 miles, park in the large university parking lot on the west side of the road. Pick up the trail on the east side.

HIGH ALPINE BREWING COMPANY

With the tagline, "Where beer meets tree line," the adventure-focused High Alpine Brewing Company opened its doors in 2015. Its beautiful, exposed-brick interior exudes coziness. Drop by for weekly live music, sample a delectable brick-oven pizza, and get ready to nurse the full lineup of taps that range from a dank pale ale made with fresh hops grown in nearby Paonia to rich, dark stouts like this one that's infused with Colorado-roasted espresso.

CONTACT INFORMATION
BLM,
Gunnison Field Office,
210 W. Spencer Avenue,
Gunnison, CO 81230;
970-642-4940

BREWERY/RESTAURANT
High Alpine Brewing Company
111 North Main Street,
Gunnison, CO 81230
970-642-4500
Miles from trailhead: 1

PONCHA SPRINGS

BAG A CLASSIC COLORADO 14ER—OR TWO

▷⋯ STARTING POINT	⋯✗ DESTINATION
BLANK GULCH TRAILHEAD	**MOUNT SHAVANO**
🍺 BEER	HIKE TYPE
APIS IV BELGIAN QUADRUPEL	**VERY STRENUOUS**
$ FEES	SEASON
NONE	**JULY TO SEPTEMBER**
⚐ MAP REFERENCE	🐾 DOG FRIENDLY
TRAILS ILLUSTRATED 130: SALIDA, ST. ELMO, MOUNT SHAVANO	**YES**
🕐 DURATION	↦ LENGTH
6-8 HOURS	**8.6 MILES**
↑↓ LOW POINT / HIGH POINT	〰 ELEVATION GAIN
9,780 FEET / 14,229 FEET	**4,461 FEET**

BELGIAN QUAD

REDDISH BLACK

BOOZY, EARTHY, MOLASSES

PLUMS, FIGS, HONEY, FLORAL

BITTERNESS IBU: 41	SWEETNESS

5
4
3
2
1

5
4
3
2
1

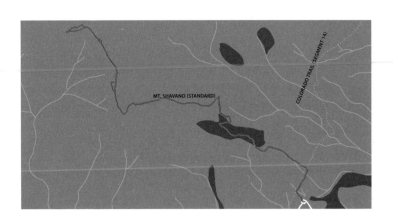

MT. SHAVANO (STANDARD)

COLORADO TRAIL SEGMENT 14)

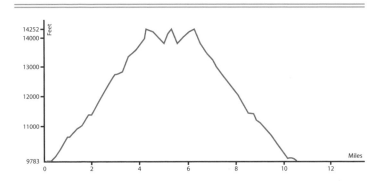

Feet

14252
14000

13000

12000

11000

9783

0   2   4   6   8   10   12

Miles

DESCRIPTION OF THE ROUTE

The most intense hike in this book deserves to be paired with the most intense beer—a classic Colorado 14er, followed by a boozy Dark Belgian-Style Ale.

The state of Colorado is home to 53 "14ers"—that is, peaks whose summits reach at least 14,000 feet above sea level, and which rise at least 300 feet in prominence above saddles connecting it to any neighboring 14ers. It seems that no Colorado hiking guidebook would be complete without at least one 14er hike in it; there is no doubt something magical about standing atop a peak that soars so close to the sky.

Many of the 14ers receive high foot traffic due to their peakbagging bragging rights. However, Shavano, and its neighboring 14er, Tabeguache, are far enough away from metropolitan Denver—and the hike to their summits just challenging enough—that they see far fewer crowds than more popular 14ers in Colorado's Front Range.

Starting at the Blank Gulch Trailhead, this hike begins on a brief stretch of the Colorado Trail, a 486-mile long-distance trail that traverses the state from Denver to Durango. You'll be on it less than a half-mile, though, before veering off onto the Shavano Trail. From here, it's a steady schlep up the mountain, winding up through beautiful, thick forests of pine and aspen trees before popping out above tree line. Here, keep an eye out for marmots, pikas, elk, and other wildlife that love to frequent the alpine tundra. Enjoy the far-reaching views of the valley below and neighboring peaks.

After you reach the saddle at mile 3.7, follow the trail to the right to scramble (Class 2) up to the rocky summit of Shavano in another half-mile or so. Enjoy the sense of accomplishment when you reach the summit. Rest and turn back here or, if you're feeling up to it and the weather is permitting, you can also tack on a second 14er summit by hiking over to nearby Tabeguache Peak (for 10.6 miles roundtrip, with 5,300 feet of total elevation gain). It is the massive rock pile to the northwest. Follow the stony ridge over to it, reaching the low saddle in a half-mile, followed by a short, Class 2 scramble to the true summit. The only way to return is to go back the same way you came, including back up and over Shavano; don't try to skirt around it, as you'll get yourself into trouble.

Even without adding Tabeguache, this hike is long and extremely strenuous, especially as the air thins as you ascend. It's best to do some altitude acclimating before attempting any 14er.

And, as with any hike that goes above tree line, this should not be attempted if lightning is present or imminent. Thunderstorms often roll through in the early afternoon, so start very early in the day and be willing to turn back shy of the summit if storms develop before you make it up there.

TURN BY TURN DIRECTIONS:

1. Start on the Colorado Trail. At 0.15 miles, turn right onto the Shavano Trail, then left at 0.4 miles.
2. At 3.7 miles, reach the saddle. Follow trail to the right to scramble up to the rocky summit.
3. At 4.3, reach the Shavano summit. Turn around here, or tack on the Tabeguache summit by following the stony ridge toward the massive rockpile to the northwest for 1 more mile.

FIND THE TRAILHEAD

From Poncha Springs, drive west on Highway 50 for about 2 miles. Make a right onto County Road 250. In 5 miles, stay straight/left to follow County Road 252 for another 3.2 miles to the Blank Gulch trailhead for Mt. Shavano/ Tabeguache Peak.

ELEVATION BEER COMPANY

Elevation Beer Company is an absolute must-visit for any self-respecting beer enthusiast traipsing across Colorado. Its cozy tasting room sits in the shadows of the giants of the Sawatch Range of the Rockies, and its beers are categorized from the mild green series (light, sessionable ales) to the moderate blue series (more traditional beers) to the knock-you-on-your-ass black and double-black series (high-ABV, barrel-aged imperial brews)—the latter of which includes this decadent, boozy Belgian quad.

CONTACT INFORMATION
U.S. Forest Service,
Salida Ranger District,
5575 Cleora Road,
Salida, CO 81201;
719-539-3591

BREWERY/RESTAURANT
Elevation Beer Company
115 Pahlone Parkway,
Poncha Springs, CO 81242
719-539-5258
Miles from trailhead: 10.8

BUENA VISTA

A FUN RAMBLE THROUGH THE DRY, SCRUB-DESERT ENVIRONS

▷⋯ STARTING POINT	⋯✕ DESTINATION
EDDYLINE BREWERY	**BROKEN BOYFRIEND TRAIL**
🍺 BEER	🎟 HIKE TYPE
JOLLY ROGER BLACK LAGER	**EASY-MODERATE**
$ FEES	📅 SEASON
NONE	**APRIL TO NOVEMBER**
⛰ MAP REFERENCE	🐾 DOG FRIENDLY
WHIPPLE TRAIL OPTIONS ON BLM.GOV	**YES (LEASH REQUIRED)**
🕐 DURATION	↦ LENGTH
1.5-2 HOURS	**4 MILES**
↑↓ LOW POINT / HIGH POINT	〰 ELEVATION GAIN
7,923 FEET / 8,392 FEET	**400 FEET**

  LAGER

 BLACK

 EARTHY, SMOKY

 CRISP, MALTY; TOFFEE UNDERTONES

BITTERNESS
IBU: 20

SWEETNESS

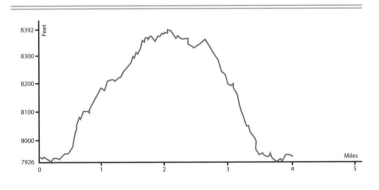

DESCRIPTION OF THE ROUTE

A fun ramble into the dry, scrub-desert environs above the outdoor enthusiasts' playground that is Buena Vista.

For an in-your-face introduction to the adventurous spirit that permeates Buena Vista, look no further than the Barbara Whipple Trail system that lords over the east side of town. These trails are neither overly crowded, nor deeply secluded. Rather, they're immersed in the heartbeat of Buena Vista's outdoors-loving community, and offer a quick way to gain elevation above town and enjoy panoramic views of the surrounding Collegiate Peaks.

Most of the trails were designed with mountain bikers in mind (so always keep an eye out for bikes on this loop), but they're a favorite haunt of hikers as well. New trails are being created each year. You can access them directly from the Eddyline Brewery by hopping on Riverside Trail, a crushed-gravel route that parallels the Arkansas River. Along your way upstream, enjoy listening to the roar of the rapids and watching for kayakers testing their mettle in the whitewater playholes.

After crossing the Whipple Bridge (also accessible directly from downtown Buena Vista, at the end of East Main Street), start climbing in earnest up the winding singletrack on sandy, hard-packed soil. You'll pass through boulder gardens, scrubby pine trees, rock outcroppings, patches of cacti, and other flowering shrubs. At 0.6 miles, take a breather on the bench to enjoy panoramic views across the valley of the Collegiate Peaks range— some of Colorado's highest "14ers" (14,000+ foot mountains). Directly below, you'll have a clear view of town.

Continue on your ramble through the foothills until you come back down to the river. Here, relatively new singletrack takes you south along the water to a different bridge than the one you first crossed. This one is on private property, but remains open to respectful public foot traffic. Cross the bridge and complete your loop by arriving back at the doorstep of Eddyline—no doubt parched and ready for imbibing.

TURN BY TURN DIRECTIONS:

1. Start by going north (upstream), following the crushed-gravel trail along the river until you reach the bridge at 0.4 miles. Cross it and continue up the singletrack.
2. At 0.5 miles, stay left at an unsigned junction.
3. At 1.1, reach a dirt road. Go left for a short stretch, then hang a right at the T6032A sign (Broken Boyfriend Trail).
4. At 2.6, turn right. Soon, you'll cross the dirt road again. Continue on the trail.
5. At the structure at 3.4, make a left onto the trail dipping into the woods.
6. At 3.8, cross the bridge across the river, then hang a right back onto the River Trail.

FIND THE TRAILHEAD

From the front step of the Eddyline Brewery, cross South Main Street and walk toward the river until you hit the Riverside Trail. Head north along the Riverside Trail for about a half mile to the Whipple Bridge.

EDDYLINE BREWERY

Established in 2009, Eddyline's aim has always been to quench the thirst of the outdoor adventure community. Emblazoned on the bartenders' shirts is their motto: "Beer for every adventure!" And the bright, playful aesthetic of the brewpub fits; bicycles, kayaks, paddles, and mountain-inspired art adorn its walls and vaulted ceiling. A perennial award winner, the rich Jolly Roger looks like a stout, but drinks smooth and easy as a lager.

CONTACT INFORMATION
BLM Colorado,
Royal Gorge Field Office,
3028 East Main Street,
Cañon City, CO 81212;
719-269-8500

BREWERY/RESTAURANT
Eddyline Brewery
926 S Main Buena Vista, CO 81211
719-966-6018
Miles from trailhead: 0.5 miles

LEADVILLE

A SECLUDED LAKE MORE THAN TWO MILES ABOVE SEA LEVEL

▷⋯ STARTING POINT	⋯✕ DESTINATION
WEST END OF TURQUOISE LAKE	**TIMBERLINE LAKE**
🍺 BEER	🎲 HIKE TYPE
PRACTICE PARADE SCOTCH ALE	**MODERATE**
$ FEES	📅 SEASON
NONE	**MARCH TO OCTOBER**
⛰ MAP REFERENCE	🐾 DOG FRIENDLY
TRAILS ILLUSTRATED 110: LEADVILLE/FAIRPLAY	**YES (LEASH REQUIRED)**
🕐 DURATION	↦ LENGTH
2-3 HOURS	**5-6 MILES**
↑↓ LOW POINT / HIGH POINT	〰 ELEVATION GAIN
10,056 FEET / 10,984 FEET	**1,050 FEET**

ALCOHOL 7.1% CONTENT

SCOTCH ALE

MAHOGANY

BURNT TOAST

SWEET MALT, CARAMEL

BITTERNESS	SWEETNESS
IBU: 30	

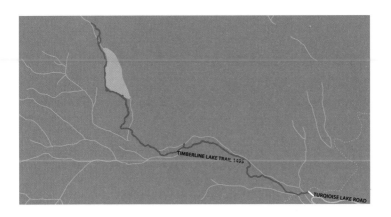

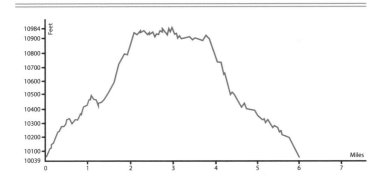

DESCRIPTION OF THE ROUTE

A relaxing hike through lush forest in the Holy Cross Wilderness area to a secluded lake more than two miles above sea level.

The highest incorporated city in North America at 10,152 feet, Leadville is known for its numerous lakes. Several of them, including Turquoise Lake and Twin Lakes, adorn the routes of Leadville's celebrated endurance races, the Leadville Trail 100 (running) and the Leadville Trail 100 MTB (mountain biking). However, beyond these well-known and much-loved landmarks lie a number of quieter, less visited lakes. One of the most pristine is Timberline Lake, nestled away in the Holy Cross Wilderness, beckoning you to escape the summer jeeping, cycling, and racing crowds. You can even camp along its shores, as a number of secluded backcountry campsites dot the water's edge.

By Colorado standards, the hike is a mellow one, meandering at a gentle grade through lush woods before emerging at the lake. The trail is rocky in places, with a number of creek crossings that will almost certainly soak your feet.

From the trailhead, this hike shares its route for the first tenth of a mile with the illustrious Colorado Trail, a 486-mile route connecting Denver to Durango. Shortly after crossing a bridge over Lake Fork Creek, you'll see your trail to Timberline Lake turn away from the Colorado Trail. Pass the two large railroad-tie Xs and, within a few minutes, enter the Holy Cross Wilderness Area. From here, the trail is easy to follow, passing through woods, marshes, rocky washes, and the occasional meadow. Through the trees you'll glimpse the surrounding high ridges, many of which stay laced with snow through the summer.

Fishing in Timberline Lake is permitted, but catch-and-release, fly-and-lure only. Take a breather or enjoy a picnic at its shores, or meander as long as you like by following an increasingly faint trail around the west side of the lake to a marshy meadow beneath an eye-popping cirque. When you're ready, turn around and head back down the way you came up.

TURN BY TURN DIRECTIONS:

1. At the immediate "fork" in the trail, stay right on the main trail to cross a bridge over the creek.
2. At 0.1 miles, stay right and follow sign for Timberline Lake.
3. At 0.3, enter the Holy Cross Wilderness Area.
4. At 2.15, reach the edge of the lake. You can follow an increasingly faint trail around the west side of the lake as long as you like.
5. Head back down the way you came up.

FIND THE TRAILHEAD

The drive to the trailhead along either the north or south sides of Turquoise Lake packs in a great deal of scenery before you even start hiking. From downtown Leadville, take County Road 4 west for 7 miles. Turn right onto Turquoise Lake Road, then find the trailhead in 2 miles at the sharp bend in the road.

PERIODIC BREWING

Calling itself the "World's Highest Craft Brewery" at an elevation of 10,156 feet, Periodic Brewing opened its doors in 2015. (For six years prior, the old mining town of Leadville had been devoid of a brewery.) Its creations span the spectrum from a light Kölsch to a hoppy IPA to this rich Scotch Ale. Their logo sports the letters "Pb," which double as an abbreviation for the brewery's initials and, cleverly, the symbol for the element of lead on the periodic table—get it?

CONTACT INFORMATION
U.S. Forest Service,
Leadville Ranger District,
810 Front Street,
Leadville, CO 80461;
719-486-0749

BREWERY/RESTAURANT
Periodic Brewing
115 E 7th St,
Leadville, CO 80461
719-270-1051
Miles from trailhead: 9.9

FAIRPLAY

VISIT AN OTHERWORLDLY GROVE OF ANCIENT TREES

▷··· STARTING POINT	···✕ DESTINATION
FOURMILE CAMPGROUND	**LIMBER GROVE**
🍺 BEER	🎲 HIKE TYPE
CHERRY BLONDE ALE	**EASY**
$ FEES	📅 SEASON
NONE	**JUNE TO SEPTEMBER**
⛰ MAP REFERENCE	🐾 DOG FRIENDLY
LIMBER GROVE BROCHURE ON MRHI.ORG	**YES (LEASH REQUIRED)**
🕐 DURATION	↦ LENGTH
1.5-2 HOURS	**3 MILES**
↑↓ LOW POINT / HIGH POINT	〰 ELEVATION GAIN
10,518 FEET / 10,938 FEET	**370 FEET**

 BLONDE ALE

 CRIMSON RED

FLORAL; BERRIES

 FIZZY, LIGHT-BODIED; SUBTLE CHERRY

BITTERNESS **SWEETNESS**
IBU: 28

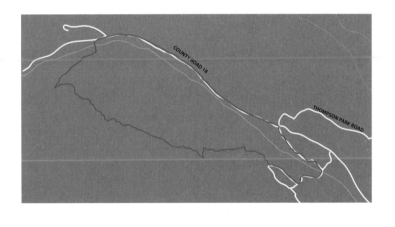

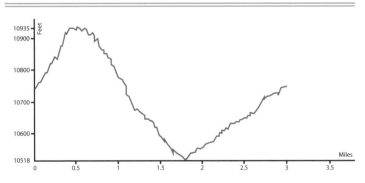

DESCRIPTION OF THE ROUTE

Unusual hikes call for unique brews—visit this otherworldly grove of ancient trees, then indulge in a spritzy blonde ale fermented with a pound of fresh cherries per gallon of beer.

The limber pine is a fascinating tree. Remarkably adapted to growing in arid, rocky environs where most other tree species cannot survive, this rare tree often grows improbably on cliff edges or in the middle of high-altitude talus slopes. This ancient limber grove just outside of Fairplay, nestled into a wind-swept pocket of talus just below 11,000 feet, is no exception.

A short hike through dense forest leads to a more open, rocky slope. A bit of careful footing across the rocks will take you to the edge of the limber grove—an outdoor museum of gnarled, twisted trees, many more than 1,000 years old.

True to their name, the limber pine's branches are flexible enough to withstand the high winds in the areas in which they thrive—so flexible, in fact, that their smaller branches can often be tied in knots without breaking. (Don't be tempted, though; this practice is discouraged by Leave No Trace principles.) It's also important not to leave the trail through the limber grove, as doing so can wear away the soil and protective bark around the limber pine's roots, eventually weakening and killing the tree.

The grove really is the highlight of this trail, so if you're just in the mood for a short hike, you can turn around here and return to the trailhead near the Fourmile Campground. If you'd like to complete a loop, though, it's also possible to continue on the trail through the woods for another mile to Horseshoe Campground. From here, you can wander through the campground out to the dirt road you drove in on and walk back up along the creek.

In early to mid-fall, especially, this makes a nice loop because the road is a corridor of fiery aspens in places, with clear views of many brilliantly bright aspen stands on the shoulders of surrounding peaks otherwise bathed in evergreens. From the road, you'll also enjoy a nice view of the 14,035-foot Mount Sherman, a lesser-known 14er—a nice option in this area if you're interested in a longer day hike.

TURN BY TURN DIRECTIONS:

1. Cross the log bridge over the creek and follow the clear singletrack.
2. At 0.5 miles, enter the limber grove.
3. At 1.5, reach Horseshoe Campground. Stay left on the road.
4. At 1.8, make a left onto the doubletrack trail, which will soon rejoin the road.
5. Go left for one more mile on the road to return to your car.

FIND THE TRAILHEAD

Head south on U.S. Route 285 for 1 mile. Turn right onto Fourmile Creek Road and drive 8 miles. Just past the turnoff for Fourmile Campground, look for the Limber Grove trailhead on your left.

SOUTH PARK BREWING

In the late 1800s, Austrian-born Leonhard Summer opened a brewery and neighboring saloon in Fairplay, which was then called South Park City. (And, yes, is the town on which the TV show, South Park, is based.) In 2014, husband-and-wife team Paul Kemp and Megan Sebastian revived the region's longtime brewing tradition by opening up their South Park microbrewery and spacious taproom. Brewed at 9,953 feet, their broad range of beers includes this crisp, dry cherry blonde, a seasonal poblano amber, a decadently rich "s'mores stout," and many other tantalizing options.

CONTACT INFORMATION
U.S. Forest Service,
South Park Ranger District,
320 Highway 285,
Fairplay, CO 80440;
719-836-2031

BREWERY/RESTAURANT
South Park Brewing
297 1/2 U.S. Route 285,
Fairplay, CO 80440
719-836-1932
Miles from trailhead: 9

BRECKENRIDGE

WOODS, WATERFALLS, LAKES, MINING RUINS, AND MORE

▷⋯ STARTING POINT	⋯✗ DESTINATION
MOHAWK LAKES 2WD TRAILHEAD	**MOHAWK LAKES**
🍺 BEER	⊞ HIKE TYPE
COCONUT PORTER	**MODERATE-STRENUOUS**
$ FEES	📅 SEASON
NONE	**JULY TO OCTOBER**
⌂ MAP REFERENCE	🐾 DOG FRIENDLY
TRAILS ILLUSTRATED 109: BRECKENRIDGE, TENNESSEE PASS	**YES**
🕐 DURATION	↦ LENGTH
4-6 HOURS	**7 MILES**
↑↓ LOW POINT / HIGH POINT	〰 ELEVATION GAIN
10,407 FEET / 12,136 FEET	**1,720 FEET**

ALCOHOL 5.9% CONTENT

PORTER

 DARK BROWN

HEAVY COCONUT AND SMOKY CARAMEL WITH A FOAMY HEAD

TOASTED COCONUT

BITTERNESS
IBU: 28

SWEETNESS

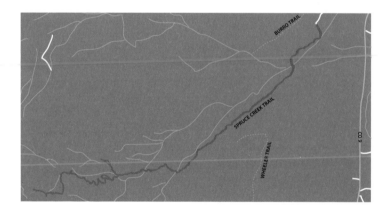

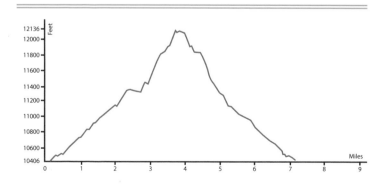

DESCRIPTION OF THE ROUTE

This hike has woods, waterfalls, wildflowers, glittering alpine lakes, historic mining ruins, postcard views, and a mix of easy ambling and gut-busting climbs.

A secluded trail this is not—but it's popular for good reason. Over the course of just a few miles, it packs in a wallop of astonishing sights and views. And basking in the sun by the shimmering water's edge of Mohawk Lakes will have you jonesin' for the rich, tropical-inspired coconut porter that awaits you in Breckenridge at the end of your hike.

Start early in the day to avoid the crowds. Almost immediately after beginning your hike at the lower trailhead, bear right at what appears to be a fork in the trail. Continue following the rocky doubletrack on its gentle climb up through the woods; be mindful of your footing, as a number of exposed roots will trip you up if you don't watch where you're going. If you start your hike early in the day, the trail is largely shaded and stays quite cool.

Shortly after reaching the upper trailhead, you'll have the option to take a short spur over to Mayflower Lake. This quiet, unassuming pond sits at the bottom of a large talus slope, and is a nice spot for a snack break. Rejoin the main trail to Mohawk Lakes and you'll quickly arrive at a large rock garden and some mining ruins, including an old log cabin.

The trail braids quite a bit here into several different social trails, most of which wind up in the same place, but if in doubt, follow cairns and/or other hikers. Bear right of the rock garden to check out the waterfall before doubling back to the main trail. A trail sign just above the log cabin will point you in the correct direction to continue on your way to Mohawk Lakes.

From here, the trail grows considerably steeper and more challenging, in a few spots you may need to use your hands for light scrambling. Pay attention to cairns again to stay on the main trail. You'll pass more historic mining elements before arriving at Lower Mohawk, a picturesque lake flanked by snow-covered cliffs, cascading streams, more old mining structures and, to the right, the looming 13,176-foot Mount Helen.

Stay left to follow the main trail to the upper Mohawk Lake after a short, steep climb with huge payoff—a cerulean lake, abundant wildflowers, high granite ridges, and commanding views of the valley behind you. Bring a jacket, as it tends to get windy up here.

TURN BY TURN DIRECTIONS:

1. From the lower 2WD trailhead, start by following the blue-diamond blazes.

2. At mile 1.5, cross the well-signed Wheeler Trail.

3. At mile 2, reach the upper 4WD trailhead. Walk to the end of the parking lot and find the Mayflower Lake trailhead sign to the right of the Spruce Creek Diversion.

4. At 2.3, you have the option to take the short spur (0.3 mile out-and-back) to Mayflower Lake.

5. Cross a log bridge over the creek and bear right on the main trail. At the rock garden, take the short spur out to the waterfall before rejoining the main trail up to Mohawk Lakes.

6. At 3.4, arrive at Lower Mohawk Lake.

7. At 3.7, arrive at Mohawk Lake.

FIND THE TRAILHEAD

Drive 3.0 miles south of downtown Breckenridge on Highway 9 and making a right on Spruce Creek Road. Follow the "Lakes & Trails" wooden signs to the lower Spruce Creek trailhead. If you have a high-clearance 4WD vehicle, you can continue up the road to an upper trailhead that shortens the entire hike substantially, but the turn by turn directions are written from the 2WD lot.

BROKEN COMPASS BREWING

Broken Compass Brewing opened its doors in 2014, quickly establishing itself as a favorite haunt of locals and visitors alike. Paying homage to Breckenridge's reputation as an epic mountain town, the tasting room is decked out with ski-lift benches and gorgeous beetle-kill wood tables and bar tops. The brew flavors range from more traditional ales to creative, visionary combinations—zingy chili pepper pale ale, peppery ginger pale ale, and this perennial favorite, the coconut porter.

CONTACT INFORMATION

U.S. Forest Service,
Dillon Ranger District,
680 Blue River Parkway,
Silverthorne, CO 80498;
970-468-5400

BREWERY/RESTAURANT

Broken Compass Brewing
68 Continental Court, Unit B12,
Breckenridge, CO 80424
970-368-2772
Miles from trailhead: 6.5

SILVERTHORNE

PART HIKE, PART SCRAMBLE IN THE EAGLES NEST WILDERNESS

▷⋯ STARTING POINT	⋯✗ DESTINATION
BUFFALO MOUNTAIN TRAILHEAD	**BUFFALO MOUNTAIN SUMMIT**
🍺 BEER	🏁 HIKE TYPE
BARKING DOG BROWN ALE (ENGLISH MILD)	**VERY STRENUOUS**
$ FEES	📅 SEASON
NONE	**JULY TO SEPTEMBER**
⛰ MAP REFERENCE	🐾 DOG FRIENDLY
TRAILS ILLUSTRATED 108: VAIL, FRISCO, DILLON	**YES (LEASH REQUIRED)**
🕐 DURATION	↦ LENGTH
5-7 HOURS	**5.3 MILES**
↕ LOW POINT / HIGH POINT	〰 ELEVATION GAIN
9,790 FEET / 12,777 FEET	**3,000 FEET**

ALCOHOL 4.8% CONTENT

BROWN ALE

 DARK BROWN, MURKY

 MALTS, TOFFEE

 CHOCOLATE, TOFFEE, MALTY

BITTERNESS IBU: UNLISTED	SWEETNESS

5
4
3
2
1

5
4
3
2
1

RYAN GULCH ROAD

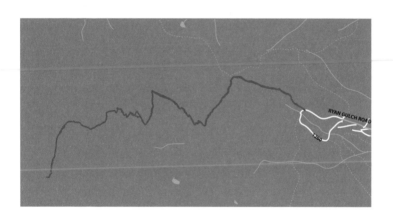

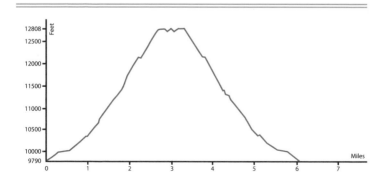

DESCRIPTION OF THE ROUTE

Bag this classic peak with a combination of hard hiking and mild scrambling, then enjoy ogling it over a brown ale from the windows of Silverthorne's Bakers' Brewery.

Getting up Buffalo Mountain is no walk in the park, but if you're up for the challenge, you'll be handsomely rewarded. The route offers fun scrambling and route finding through the talus slopes, bird's-eye views of the Dillon Reservoir and nearby Grays and Torreys peaks, and a chance to see the herd of mountain goats that frequents the upper slopes of this mountain. Plus, after a hike like this one, the brown ale that awaits you at Brewers' Bakery will taste all the more magnificent.

The hike begins at a relatively mellow grade, passing old mining ruins and subalpine forest on the way into the Eagles Nest Wilderness. After the first mile, you'll begin encountering sizable talus slopes. Be mindful to follow the cairns to guide you through any areas in which the "trail" becomes difficult to find. Hundreds of cairns exist in these boulder fields, easing your way to the summit with an established climbers' route. Nonetheless, you're likely to need both hands at times to help you scramble over certain sections. This trail is not for the faint of heart.

As you climb, enjoy views of the Dillon Reservoir behind you. At just over two miles, you'll emerge from the largest talus slope at the saddle of a false summit. Next up is another, slightly smaller talus slope and more cairn-following before you emerge on the ridgeline. The true summit is here—so if you're ready, make yourself at home, layer up against the wind, and enjoy the panoramic views. Keep an eye out for mountain goats, which love to roam the alpine meadows here; you'll no doubt find clumps of their white hair clinging to rocks along the summit.

For the bravest of souls, there is an optional scramble with some moderate exposure to a secondary summit farther along the ridgeline.

Start this hike early in the day, check the forecast to avoid thunderstorms, and budget lots of time; picking your way through the rock gardens and talus means mandatorily slow going for much of the way. Also, even on a hot summer day, be sure to pack a jacket, warm cap, and gloves for the brisk winds you're likely to encounter at the summit.

TURN BY TURN DIRECTIONS:

1. At 0.57 miles, bear left at the four-way trail intersection.
2. At 1.15, start climbing steeply through the talus. Look for cairns to guide you and be mindful not to get off the main route.
3. At 2.2, emerge at the saddle of a false summit.
4. At 2.6, arrive at another talus slope. Follow cairns again.
5. At 2.8, emerge on the ridgeline and summit. An optional scramble (with some exposure) across the ridge takes you to a secondary summit at just over 3 miles.

FIND THE TRAILHEAD

Follow Wildernest Road south from Silverthorne, then west until it turns into Ryan Gulch Road. The well-signed trailhead sits at the top (west end) of this road.

THE BAKERS' BREWERY

This bakery-brewery combination is the masterpiece of Cory Forster, former head brewmaster at the nearby Dillon Dam Brewery, one of the largest brewpubs in the country. Unlike many establishments that have either great beer but mediocre food or great food but mediocre beer, The Bakers' Brewery nails both fronts, with fresh housemade sourdough bread and solid brews ranging from sessionable to boldly experimental. Impressive views of Buffalo Mountain treat those who snag the window seats.

CONTACT INFORMATION
U.S. Forest Service,
Dillon Ranger District,
680 Blue River Parkway
Silverthorne, CO 80498;
970-468-5400

BREWERY/RESTAURANT
The Bakers' Brewery
531 Silverthorne Lane,
Silverthorne, CO 80498
970-468-0170
Miles from trailhead: 3.4

EDWARDS

AN EASY, QUIET STROLL THROUGH TOWERING ASPEN GROVES

▷⋯ STARTING POINT	⋯✕ DESTINATION
END OF W. LAKE CREEK DRIVE	**BRIDGE OR BOOT LAKE**
🍺 BEER	🎛 HIKE TYPE
VAIL TAIL PALE ALE	**MODERATE**
$ FEES	📅 SEASON
NONE	**MAY TO OCTOBER**
⛰ MAP REFERENCE	🐾 DOG FRIENDLY
TRAILS ILLUSTRATED 149: EAGLES NEST AND HOLY CROSS WILDERNESS AREAS	**YES (LEASH REQUIRED)**
⏱ DURATION	↦ LENGTH
3-8 HOURS	**4-12 MILES**
↑↓ LOW POINT / HIGH POINT	〰 ELEVATION GAIN
8,294 FEET / 9,636 FEET	**2,090 FEET**

PALE ALE

DEEP GOLD

FLORAL, CITRUS

MALTY

BITTERNESS	SWEETNESS
IBU: 78	

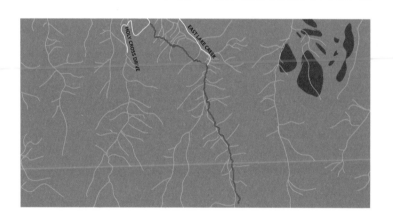

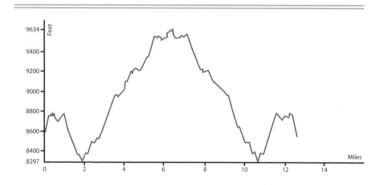

DESCRIPTION OF THE ROUTE

An easy, quiet stroll through towering aspen groves and deep spruce and fir forest, followed up with hops galore.

It can be hard to escape the crowds along the I-70 corridor, but this gem of a trail just southwest of Vail is a lightly trafficked respite from it all. The East Lake Creek Trail stretches nearly 13 miles one way, largely in the Holy Cross Wilderness, so it can be walked as an out-and-back of any distance. The bridge at 2.2 miles in makes for a popular turnaround spot for a shorter hike, or you can opt for the longer, more forested stroll to Boot Lake a little over 6 miles in. ("Lake" is a generous word to describe this marshy pond, though.)

There are few landmarks along the way as the trail gradually wends its way up the belly of a deep valley—so the pleasure of this trail, especially for a reasonable day hike, lies not in panoramic views but in the solitude and quiet of tromping through remote woods. Mushroom hunters, berry pickers, birdwatchers, wildflower enthusiasts, or autumn leaf peepers will revel in the valley's abundant flora and fauna. The gurgle of the creek—which parallels much of the trail—and its occasional gentle cascades offer a nice soundtrack to your hike. Finally, the colossal aspen groves along the first three miles of the trail make for spectacular colors every fall, as the leaves turn to blazing golden hues.

The first stretch passes through private property on its way toward the Holy Cross Wilderness boundary, so be sure to stick to the trail here. You'll start with a short, stout climb up through towering aspen trees before it gives way to a mellower, rolling hillside traverse. The trail can be narrow and overgrown in places—be prepared for lots thimbleberry bushes and ferns brushing up against your legs—but it's never difficult to follow.

The first few miles afford the most expansive views of the valley and surrounding high ridgelines, as well as the most impressive aspen groves. At 2.2 miles you'll come to the major bridge crossing—a popular lunch spot or turnaround for a shorter day hike. About half a mile later, the trail sidles up next to the creek and turns distinctly lusher and rockier, scattered with spruce and fir needles. For miles to come, it stays well shaded.

At 6.1 miles, the valley opens up a bit again, with a large grassy meadow (and the tiny Boot Lake) to your right, and a large, boulder-strewn avalanche chute to your left. Keep an ear open for the whistle of marmots greeting you as you arrive to their rocky playground. You can hike as little or as far as you like before turning around, but heading back here will allow you to get back to Edwards in time to soak up the final hours of daylight on Gore Range Brewery's outdoor patio.

TURN BY TURN DIRECTIONS:

1. At 0.8 miles, pass the turnoff for Dead Dog Trail. Stay left/straight.
2. At 1.8, enter the Holy Cross Wilderness Area.
3. At 2.2, come to a bridge over the creek. This is a popular turnaround spot for a shorter day hike.
4. At 6.1, tiny Boot Lake sits in the meadow to your right. You can soldier on for as far as you like on this trail, or call it a day here and head back the way you came.

FIND THE TRAILHEAD

From downtown Edwards, drive west on U.S. Route 6. In less than a mile, turn left onto Lake Creek Road. In 1.8 miles, turn right onto W Lake Creek Road and follow that for 2.6 miles before making a sharp left onto W Lake Creek Drive. Arrive at East Creek Lake trailhead in less than a mile.

GORE RANGE BREWERY

Gore Range Brewery's history in the Vail Valley dates back to 1997, but it passed to new ownership in 2011—chef Pascal Coudoy—and adopted a new head brewer, Jeff Atencio, in 2016. By design, the beers are light, easy-on-the-palate choices made to accompany Coudoy's burgers, wood-fired pizzas, grill specialties and other American-style pub fare. They don't distribute their handcrafted beer anywhere, so pulling up a seat at their family-friendly restaurant is the only way to sample it. For fans of easy-drinking ales, the Vail Tail Pale Ale is a 95% gluten-reduced beer that's brewed all summer long.

CONTACT INFORMATION
U.S. Forest Service,
Eagle Ranger District,
24747 US Highway 24,
Minturn, CO 81657;
970-827-5715

BREWERY/RESTAURANT
Gore Range Brewery
105 Edwards Village Blvd,
Edwards, CO 81632
Miles from trailhead: 6.1
970-926-2739

EAGLE

**ENJOY SUNRISE OR SUNSET IN THESE FOOTHILLS
ON THE FRINGES OF TOWN**

▷⋯ STARTING POINT	⋯✕ DESTINATION
POLAR STAR DRIVE	**REDNECK RIDGE**
🍺 BEER	HIKE TYPE
KINDLER PALE ALE	**MODERATE**
$ FEES	SEASON
NONE	**APRIL TO OCTOBER**
⌂ MAP REFERENCE	🐾 DOG FRIENDLY
MOUNTAINBIKEEAGLE.COM	**YES**
⏱ DURATION	↦ LENGTH
2-3 HOURS	**5.9 MILES**
↑↓ LOW POINT / HIGH POINT	〰 ELEVATION GAIN
6,785 FEET / 7,785 FEET	**1,033 FEET**

PALE ALE

 DARK GOLDEN

 BREAD, HOPS, GRAPEFRUIT

 CRISP; CITRUS HOPS

BITTERNESS
IBU: 48

SWEETNESS

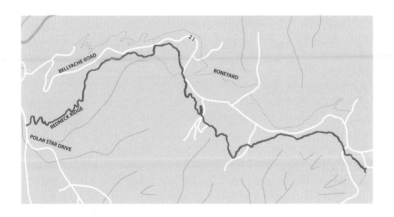

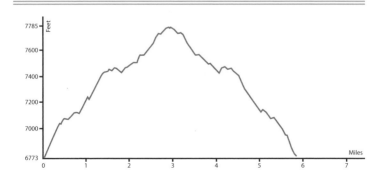

DESCRIPTION OF THE ROUTE

Take advantage of this singletrack climb up a ridge on the eastern fringes of town before sampling up to two dozen fresh beers on tap at Bonfire Brewing.

Sandwiched between Vail and Glenwood Springs on the western slope of the Rockies, Eagle is an outdoor enthusiast's paradise that rarely receives the attention it deserves. Some 300 miles of trails offer endless recreation opportunities on BLM land and national forest land within close range of downtown. The Eagle River on the north edge of town offers ample rafting and fishing opportunities, and nearby Sylvan Lake State Park is outfitted with campsites, yurts, trails and more—the perfect adventure base camp for hikers, campers, paddlers, and anglers alike. Indeed, for those willing to drive a bit farther to trailheads, the hiking opportunities in the vast tracts of national forest and wilderness to the south are truly endless.

Even closer to town, though, lies a plethora of singletrack trails for hiking, mountain biking, trail running, and equestrian use. Among them, on the eastern flanks of town, is a spider web of mountain-biking trails bearing spunky names like the Boneyard, Bellyache, Pool and Ice Rink, and Redneck Ridge.

This hike begins on chalky singletrack directly out of a residential neighborhood. At first, you'll be able to hear the buzz of the highway below, but the higher you hike, you more the sounds of traffic and civilization fade away. Though some stretches of the hike are lightly wooded with juniper and pinyon pine, much of this hike is exposed, traversing high-desert environs replete with fragrant sage, rabbitbrush, and cacti. It's a great place to soak up some early morning or late afternoon sunshine. As the trail mazes its way out of the sagebrush meadow and into more shaded juniper forest, enjoy glimpses on the horizon of the surrounding red cliffs and larger mountain ranges farther afield. Your final destination is a lookout point that affords the farthest-reaching views of all; it's just past the Pool and Ice turnoff and a campsite. Take it all in before returning the way you came and making your way back into town for a barstool inside Bonfire's abundantly stocked tasting room, or a seat around its roaring fire pit outside.

TURN BY TURN DIRECTIONS:

1. Reach an unmarked fork at 1 mile. Either way is fine; they'll meet back up in another third of a mile.
2. At 2.75, go straight through the trail junction.
3. At 2.9, reach a campsite and viewpoint. This is your turnaround.

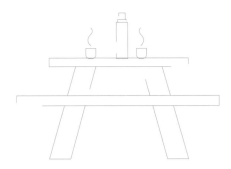

FIND THE TRAILHEAD

Head east out of town on E 3rd Street, which turns into Polar Star Drive. After rounding the first curve, look for singletrack on your left-hand side. This is your trail. If you reach the fork between Kaibab Road and Polar Star Drive, you've gone too far.

BONFIRE BREWING

Bonfire Brewing is a true neighborhood joint—a dog-friendly taproom packed with jovial patrons, foosball and shuffleboard tables, and an outdoor patio and fire pit. There's free popcorn to keep your thirst up for the astonishing number of beers on tap—often more than 20 at a time! Bring your own food, or order some local pizza in on the "bat phone" on the wall. Like many a Colorado craft brewery, Bonfire was started by a couple of home brewers eager to trade cubicle life for something better.

CONTACT INFORMATION
BLM,
Colorado River Valley Field Office,
2300 River Frontage Road,
Silt, CO 81652;
970-876-9000

BREWERY/RESTAURANT
Bonfire Brewing
127 W 2nd Street,
Eagle, CO 81631
970-306-7113
Miles from trailhead: 0.9

STEAMBOAT SPRINGS FISH CREEK FALLS

TWO MAGNIFICENT WATERFALLS FOR THE PRICE OF ONE

▷⋯ STARTING POINT	⋯✗ DESTINATION
FISH CREEK FALLS TRAILHEAD	**UPPER FISH CREEK FALLS**
🍺 BEER	🎴 HIKE TYPE
BAVARIAN HEFEWEIZEN	**MODERATE-STRENUOUS**
$ FEES	📅 SEASON
$5	**JUNE TO OCTOBER**
🗺 MAP REFERENCE	🐾 DOG FRIENDLY
TRAILS ILLUSTRATED 118: STEAMBOAT SPRINGS, RABBIT EARS PASS	**YES (LEASH REQUIRED)**
🕐 DURATION	↦ LENGTH
3 HOURS	**4.9 MILES**
↑↓ LOW POINT / HIGH POINT	〰 ELEVATION GAIN
7,440 FEET / 8,780 FEET	**1,470 FEET**

HEFEWEIZEN

ALCOHOL CONTENT 5.4%

 CLOUDY, GOLDEN

 BANANA, CLOVES; YEASTY

 CREAMY; BUBBLEGUM ESTERS

BITTERNESS
IBU: 15

SWEETNESS

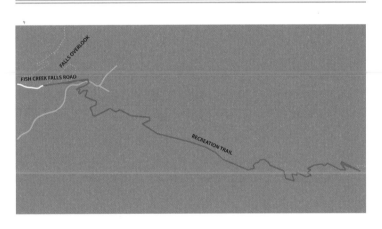

FALLS OVERLOOK

FISH CREEK FALLS ROAD

RECREATION TRAIL

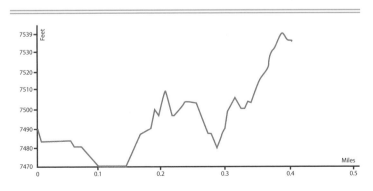

DESCRIPTION OF THE ROUTE

Visit two magnificent waterfalls in a single hike before stopping in for a Bavarian-style hef at the Boat's original craft brewery.

Steamboat Springs is a town famous for its world-class skiing, extensive network of singletrack trails, abundance of geothermal hot springs and—yes—good beer. What's not to love? The town derives its name from the chugging sound that early explorers heard and assumed was a steamboat coming down the Yampa River—only to discover that the gurgling was, in fact, coming from the natural hot springs near the old train depot.

Fish Creek Falls is one of the most popular stops for visitors to Steamboat, especially in the late spring when the falls are roaring at their fullest and early-mid fall when the aspen leaves are changing. You can avoid the "strenuous" part of this hike altogether and head (nearly) straight for the taproom at Butcherknife by keeping your hiking adventure to the main Fish Creek Falls overlook, which makes for a scenic roundtrip hike of just half a mile—downhill on your way out, uphill on your way back. The bridge at the base of the 280-foot falls offers a lovely view from afar of the cascade, as well as the babbling creek into which they spill. Visit in mid-late spring to see the falls at peak runoff.

If you're game for a longer adventure and a way to escape the crowds, continue your hike beyond the bridge to a second waterfall higher up. Begin with switchbacks that climb up through the dense, shaded forest. You'll pass through some stands of aspen along the way, which turn to fiery hues come late September. After passing a second bridge across Fish Creek, the trail grows rockier and steeper yet. From the bridge, it's just over half a mile to reach the upper falls. Enjoy the views of the valley as they open up along this final stretch of the hike. The upper falls, though slightly smaller than the lower falls, are no less impressive. They're surrounded by rocks and boulders for scrambling up and relaxing on to eat a snack before heading back the way you came for noshing and a brewski back in town.

TURN BY TURN DIRECTIONS:

1. At 0.2 miles, reach the bridge overlooking the falls.
2. Continue hiking past the bridge, crossing another bridge at 1.5 miles.
3. Reach upper falls in just over 2 miles.

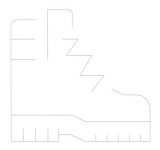

FIND THE TRAILHEAD

Head east out of downtown Steamboat Springs on Oak Street, which turns into Fish Creek Falls Road. In 2.4 miles, follow the road around in a sharp right to stay on Fish Creek Falls Road and arrive at trailhead in another 0.7 miles.

BUTCHERKNIFE BREWING COMPANY

Butcherknife Brewing Company was Steamboat's first production brewery and, to this day, prides itself on bold, experimental brews. Past innovations have included a dry-hopped, 10.1% ABV double IPA, a bourbon-barrel-aged mint-julep saison, and a porter infused with roasted coffee beans from Steamboat Coffee + Tea Company. With a gorgeous outdoor patio and beautiful, butcherknife-themed flight boards and tap pulls, it's just the kind of place that will make you want to linger.

CONTACT INFORMATION

U.S. Forest Service,
Hahns Peak/Bears Ears Ranger District,
925 Weiss Dr.,
Steamboat Springs, CO 80487;
970-870-2299

BREWERY/RESTAURANT

Butcherknife Brewing Company
2875 Elk River Road,
Steamboat Springs, CO 80487
970-879-2337
Miles from trailhead: 5.7

STEAMBOAT SPRINGS
MAD CREEK

SUNNY RAMBLE THROUGH A HISTORIC VALLEY

▷⋯ STARTING POINT	⋯✕ DESTINATION
MAD CREEK TRAILHEAD	**MAD CREEK BARN**
🍺 BEER	HIKE TYPE
MAD CREEK KÖLSCH	**EASY-MODERATE**
$ FEES	SEASON
NONE	**APRIL TO NOVEMBER**
⌂ MAP REFERENCE	🐾 DOG FRIENDLY
TRAILS ILLUSTRATED 117: CLARK, BUFFALO PASS	**YES**
⌚ DURATION	↦ LENGTH
2-3 HOURS	**4-6 MILES**
↑↓ LOW POINT / HIGH POINT	～ ELEVATION GAIN
6,772 FEET / 7,326 FEET	**550 FEET**

 KÖLSCH

 PALE GOLD

 STRAW; GRAINY

 CRISP; DELICATE MALTS

BITTERNESS	SWEETNESS
IBU: 22	

MAD CREEK
COLORADO STYLE KOLSCH

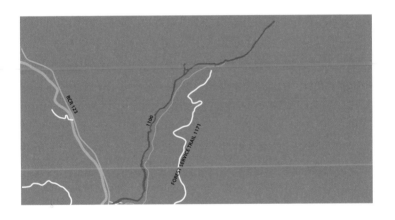

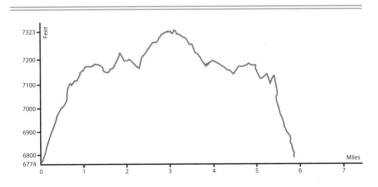

DESCRIPTION OF THE ROUTE

This sunny ramble through a historic valley calls for treating your toes to a dip in Mad Creek and your taste buds to a refreshing Kölsch named in honor of this trail.

This classic hike just outside of Steamboat Springs has a lot to offer—relatively easy walking, expansive views, aspen groves, a picturesque creek perfect for picnicking, and a few historic curiosities along the way, including the century-year-old Mad Creek Barn. This hike is sunny and exposed, so wear sunscreen and a hat, bring plenty of water, and watch out for rattlesnakes.

The biggest climb of the hike comes first. It's a manageable grade, though, up a wide horse trail. As you hike, you'll make your way into the Mad Creek Valley and toward the expansive Mount Zirkel Wilderness Area, one of the country's original wilderness areas designated by Congress in 1964.

Enjoy the gurgle of the creek below to your right as you traverse the hillside and enter a short, forested stretch. Before long, you'll pop back out of the trees on an easy, flat path. In the summer, wildflowers dot these expansive meadows. Come fall, the surrounding hillsides are blanketed in fiery scrub oak and vivid, yellow aspen leaves.

Along the way, you'll visit the Mad Creek Barn, built in 1906 by local rancher James "Harry" Ratliff, the first Forest Supervisor of the Routt National Forest. Setting foot inside the barn is like stepping back in time. Enjoy a stop here before rejoining the trail and venturing on. Shortly, you'll arrive at a bridge over Mad Creek, a peaceful lunch spot and the perfect place to dip your toes in the icy water. The remains of several old miners' cabins can be found in the woods on both sides of the creek, just past the bridge.

Turn around here, or continue further into the valley on the trail on the same side of Mad Creek that you've been hiking. Theoretically (with several days to devote to the effort), you could hike from here all the way to the Wyoming border. However, a crisp Kölsch named in honor of this hike awaits you back in Steamboat. A gorgeous aspen grove around 3 miles in makes as good a turnaround spot as any for thirsty day hikers.

TURN BY TURN DIRECTIONS:

1. At 1.7 miles, you'll see Mad Creek Barn. Veer left to visit it.
2. Rejoin the original trail and continue further into the valley.
3. Within a quarter-mile, check out the miners' cabin ruins and the bridge over Mad Creek.
4. At 3, a gorgeous swath of aspen trees makes a good turnaround spot.

FIND THE TRAILHEAD

Drive west out of Steamboat Springs on Highway 40. Storm Peak Brewing will be on your right shortly before the turnoff for Elk River Road (CO Road 129). The trailhead, well marked from the highway, is about 5.5 miles up this road.

STORM PEAK BREWING COMPANY

Owned and operated by a self-described "family of beer freaks, music lovers, and outdoor enthusiasts," Storm Peak Brewing sports a seven-barrel brew house and casual, garage-style taproom. It specializes in creative, small-batch brews, so regardless of whether this refreshing Kölsch is on tap when you visit, treat yourself to a taster flight of whichever of the rotating options are fresh in the kegs.

CONTACT INFORMATION
U.S. Forest Service,
Hahns Peak/Bears Ears Ranger District,
925 Weiss Drive,
Steamboat Springs, CO 80487;
970-870-2299

BREWERY/RESTAURANT
Storm Peak Brewing Company
1885 Elk River Plaza
Steamboat Springs, CO 80487
970-879-1999
Miles from trailhead: 5.5

IDAHO SPRINGS

PAY A VISIT TO ONE OF COLORADO'S LAST REMAINING GLACIERS

▷⋯ STARTING POINT	⋯✗ DESTINATION
ST. MARY'S GLACIER TRAILHEAD	**ST. MARY'S LAKE AND GLACIER**
🍺 BEER	🎫 HIKE TYPE
BUTT HEAD BOCK LAGER	**MODERATE**
$ FEES	📅 SEASON
$5	**JUNE TO OCTOBER (OR SNOWSHOE YEAR-ROUND)**
⛰ MAP REFERENCE	🐾 DOG FRIENDLY
TRAILS ILLUSTRATED 103: WINTER PARK, CENTRAL CITY, ROLLINS PASS	**YES (LEASH REQUIRED)**
🕐 DURATION	↦ LENGTH
1 HOUR	**2 MILES**
↑↓ LOW POINT / HIGH POINT	〰 ELEVATION GAIN
10,407 FEET / 11,207 FEET	**735 FEET**

BOCK LAGER

 DARK COPPER BROWN

 SWEET MALT, RAISINS

 TOFFEE, DARK FRUITS, TOASTED CARAMEL

BITTERNESS	SWEETNESS
IBU: 30	

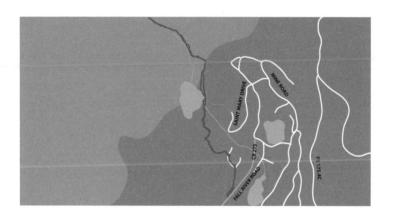

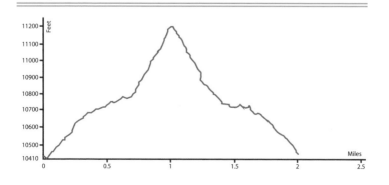

DESCRIPTION OF THE ROUTE

Visit one of Colorado's few remaining glaciers on this rewarding hike that delivers you to the foot of a lake, waterfall, and small glacier in less than a mile.

To date, Colorado has just 14 officially named glaciers remaining, most of which lie within the boundaries of Rocky Mountain National Park, or in or near the Indian Peaks Wilderness. Tiny St. Mary's lies just above an eponymous lake on the cusp of private property, but a 0.75-mile trail is open to the public for those who wish to venture to its flanks. Rewarding you for your efforts on the rocky trail up are terrific views of the alpine lake, cascading glacier, and small waterfall between them, as well as the rocky slopes and cliffs towering over the lake.

In all months of the year, locals flock to St. Mary's for the chance to ski, snowboard, sled, or glissade, even if only for a minute or two on the short snowfield—hiking up to earn their turns even at the height of summer or fall. (Do beware that dangers exist year-round, ranging from precarious snow bridges in the summer to avalanche conditions in the winter; as signs as the trailhead will warn, there have been fatalities here in the past.)

Due to this hike's scenic destination, relative brevity, and its proximity to Denver, it gets busy by midday, especially on weekends. Plan to arrive early for a better shot at easy parking and a less crowded experience at the glacier. Hiking up above the lake and glacier also offers more solitude at any time of the day, as most hikers don't venture above the bottom of the glacier.

From the trailhead, follow the wide, rock- and boulder-strewn dirt road up. Several trails and road spurs splinter off in different directions, but the abundant private property signs tacked on to trees lining the trail should keep you on track. Within a half-mile, the lake, glacier, and waterfall will be within view, along with a beautiful grove of twisted old pines and the looming nearby Fox Mountain.

From here, it's largely a matter of choosing your own adventure on the various splintering trails around the lake and up near the glacier. Beyond what's visible from the bottom, there is a whole other section of glacier—not visible from the lake—if you venture up another quarter- or half-mile. Watch your footing on the loose earth if you opt to scramble up this way. A rough path snakes up along the northwest (looker's right) side of the glacier. Follow this up as high as you like, to the top of the glacier or even beyond to a higher pass with views toward the James Peak Wilderness Area.

When you're ready to call it a day, head back down the way you came up. A malty bock lager awaits you at Tommyknocker Brewery on the main drag in Idaho Springs.

TURN BY TURN DIRECTIONS:

1. At the first two wide junctions, stay left on the main road, continuing up.
2. At 0.3 miles, follow the trail to the right.
3. At 0.5, cross the bridge and arrive at St. Mary's Lake.
4. At 0.75, arrive at the foot of St. Mary's glacier. Turn around here, or continue up the trail along the northeast side of the glacier another quarter-mile or more for more glacier views.

FIND THE TRAILHEAD

Take exit 238 off I-70 and follow signs for the glacier, traveling north on Fall River Road for about 9 miles. Bring $5 (exact bills in cash) to park in either trailhead lot; both sit on private property. The well-marked trailhead sits between them.

TOMMYKNOCKER BREWERY

Tommyknocker Brewery has been brewing lagers and ales in the historic mining town of Idaho Springs since 1994. It's earned 17 Great American Beer Festival (GABF) medals over the years, including multiple wins for this boozy, Munich-style doppelbock lager. Hearty, Rocky-Mountain-themed pub fare (think elk burgers and Butt Head Bock brats) and a full array of house-made root beers and cream sodas for the non-beer drinkers let you fuel up for all your mountain adventures.

CONTACT INFORMATION
U.S. Forest Service,
Clear Creek Ranger District,
2060 Miner Street,
Idaho Springs, CO 80452;
303-567-4382

BREWERY/RESTAURANT
Tommyknocker Brewery
1401 Miner Street,
Idaho Springs, CO 80452
303-567-2688
Miles from trailhead: 11.3

WINTER PARK

A PRETTY, LESSER-KNOWN PATH THROUGH ARAPAHO
NATIONAL FOREST

▷⋯ STARTING POINT	⋯✗ DESTINATION
RIDGE TRAILHEAD	**IDLEWILD TRAIL SYSTEM**
🍺 BEER	🀫 HIKE TYPE
CHOO CHOO CHAI MILK STOUT	**EASY**
$ FEES	📅 SEASON
NONE	**JUNE TO OCTOBER (OR SNOWSHOE/ XC SKI YEAR-ROUND)**
⛰ MAP REFERENCE	🐾 DOG FRIENDLY
IDLEWILD TRAIL MAP AT RENDEZVOUSCOLORADO.COM	**YES**
🕐 DURATION	↦ LENGTH
1.5 HOURS	**3.4 MILES**
↕ LOW POINT / HIGH POINT	〰 ELEVATION GAIN
9,012 FEET / 9,203 FEET	**300 FEET**

MILK STOUT

ALCOHOL
6.4%
CONTENT

 PITCH BLACK

 MAPLE, CLOVES, CARDAMOM

 CREAMY, CHAI SPICE, ROASTY MALTS, PEPPERY

BITTERNESS
IBU: 10

SWEETNESS

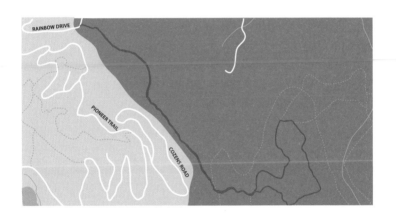

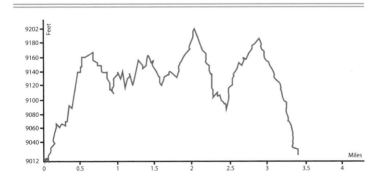

DESCRIPTION OF THE ROUTE

Take this pretty, lesser-known path through lodgepole pine and aspen to access unlimited loop options on wooded singletrack in the Arapaho National Forest.

The town of Winter Park sits just north of its eponymous ski area, Colorado's longest continually running ski resort, which opened to the public in 1939 with a single J-bar tow and one-dollar lift tickets. While it continues today to be a popular destination for skiers and snowboarders in the snowy months, come summer, the trails that surround both the ski area and the town are just as much a mecca for hikers, trail runners, and mountain bikers.

The Idlewild trail system to the northeast of town ventures into the Arapaho National Forest. Some two dozen singletrack trails snake through the dense forest. Your best bet for a great day hike here is to print off the trail map at www.rendezvouscolorado.com and use it to navigate; though trail intersections are well signed, it's easy to get confused and lost if you don't have the bigger picture of how all the trails connect.

Not pictured on that map or many others, however, is a beautiful, mellow, mile-long footpath called the Ridge Trail. It traverses private property north of the Idlewild system, but at least at the time of this book's writing, the landowner permits public access to this trail, so long as trail users do not stray off of it until they reach the well-signed Idlewild trails.

This trail is a lovely way to reach the Idlewild system, for it passes through gorgeous lodgepole pines and aspen trees, as well as past some old cabin ruins. Once you reach the other trails, infinite loop options exist. A simple, short lollipop option is detailed in the turn by turn directions here, using the South Fork Loop, Winterwoods, and Crosstrails paths before connecting back to the Ridge Trail.

Other, longer jaunts include following signs for the South Fork Loop the whole way (adds about 2 miles), or adding the lollipop Burnout Loop (about 2 additional miles) off of the northeast side of the South Fork Loop trail. Regardless, your way out is the way you came in. Then, a short drive down the hill brings you to the doorstep of Hideaway Park Brewery for a decadent, chai-flavored milk stout.

TURN BY TURN DIRECTIONS:

1. Note that this trail lies on private property, so heed any and all notices at the trailhead before proceeding through the gate. Stay on the trail.
2. At just under a mile, arrive at the Idlewild trails. With a map, you can explore here to your heart's delight.
3. Otherwise, follow signs for the South Fork Loop, staying right, then straight, then left.
4. At 1.5 miles, make a right toward the Winterwoods Trail, then a quick right again.
5. At 2.1, turn right onto Crosstrails, which will loop you back toward the first series of intersections with the Ridge Trail. Return the way you came.

FIND THE TRAILHEAD

From downtown Winter Park, drive north on Highway 40 for about two miles. Turn right on Meadow Ridge Road, right on Cranmer Avenue, right on Brooky Drive, then right on Rainbow Drive. The trailhead is at the corner; park on the side of the road.

HIDEAWAY PARK BREWERY

Hideaway Park Brewery is nestled beneath a snowboard shop and next door to a gourmet hot-dog shop on the main drag in Winter Park, just around the corner from Hideaway Park. It got its start the way so many small-town craft breweries do in Colorado—with a Denverite burnt out on the corporate rat race, ready to move to the mountains and bring craft beer to a town previously lacking it. Beers brewed in house frequently rotate, so be sure to sample whatever's fresh on tap.

CONTACT INFORMATION
U.S. Forest Service,
Sulphur Ranger District,
9 Ten Mile Drive,
Granby, CO 80446;
970-887-4100

BREWERY/RESTAURANT
Hideaway Park Brewery
78927 U.S. 40,
Winter Park, CO 80446
970-363-7312
Miles from trailhead: 4.2

NEDERLAND

STUNNING BACKCOUNTRY LAKES IN THE INDIAN PEAKS WILDERNESS AREA

▷⋯ STARTING POINT	⋯✕ DESTINATION
HESSIE TRAILHEAD 2WD PARKING	**WOODLAND LAKE AND SKYSCRAPER RESERVOIR**
🍺 BEER	🀲 HIKE TYPE
VERY NICE PALE ALE	**MODERATE-STRENUOUS**
$ FEES	📅 SEASON
NONE	**JULY TO SEPTEMBER**
⛰ MAP REFERENCE	🐾 DOG FRIENDLY
TRAILS ILLUSTRATED 103: WINTER PARK, CENTRAL CITY, ROLLINS PASS	**YES (LEASH REQUIRED)**
🕐 DURATION	↦ LENGTH
4-7 HOURS	**10.6 MILES**
↑↓ LOW POINT / HIGH POINT	〰 ELEVATION GAIN
9,012 FEET / 11,260 FEET	**2,180 FEET**

PALE ALE

ALCOHOL 5.9% CONTENT

VERY NICE
PALE ALE

1 PINT 6 FL OZ
5.9% ALC BY VOL

VERY NICE
Brewing Company

BRIGHT STRAW

FLORAL; LEMON

HOPPY, TANGY

BITTERNESS	SWEETNESS
IBU: 33	

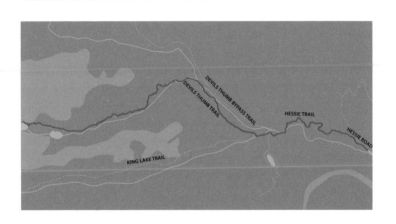

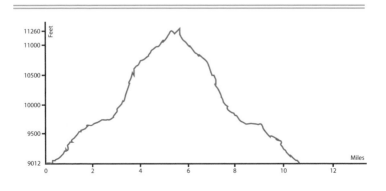

DESCRIPTION OF THE ROUTE

The long hike in is worth it for a glimpse of these backcountry gems, let alone the refreshing pale ale that awaits you at Very Nice Brewing.

It's hard to beat the views at these scenic backcountry lakes, which are tucked into impressive, craggy cirques deep inside the Indian Peaks Wilderness Area. They're located at the foot of the Continental Divide. The entire area is pockmarked with dozens of picturesque bodies of water including Devil's Thumb, King, Bob, Betty, and Storm Lakes, not to mention Woodland Lake and Skyscraper Reservoir, your destinations on this particular trek.

Once upon a time, this region was a hotbed for mining, including the successful Fourth of July Mine. The ghost town of Hessie—which you'll pass early on your journey, and from which the trailhead gleans its name—served as a homesteading site for a number of silver miners in the 1890s and early 1900s. It drew its name from the wife of town founder Captain J.H. Davis, who set up a post office and became the town's first postmistress.

Today, the area is exceptionally popular for hikers and backpackers eager to soak up its breathtaking alpine views. That said, the majority of day hikers do not venture all the way to the lakes, most of which require several hours to reach. If you're game for a longer hike, it's possible to out-hike the crowds, especially if you take advantage of visiting on weekdays or in the fall off-season.

From the 2WD parking area along the sides of the road, your hike begins on a rocky jeep road for a half-mile before arriving at a bridge and the official Hessie trailhead. The hike in to Woodland Lake has a fairly mellow grade, with a few steep pitches here and there. The lake will be on your left, visible through the trees. If you opt to continue on up to Skyscraper Reservoir (absolutely worth doing if you have the time and energy for the bonus mile!), you'll enjoy even better views of Woodland Lake from above, where carpets of wildflowers also await.

Skyscraper Reservoir is appropriately named, as its glittering waters seem to kiss the blue sky beneath the Divide. Originally built in the 1940s for irrigation purposes for farmers and ranchers downstream, the reservoir is no longer maintained; however, it retains remnants of its original damming. Lots of beautiful driftwood graces its shores, with many rocks to scramble over on any further, off-trail meandering you want to do around the lake before hiking back down the same way you came up.

TURN BY TURN DIRECTIONS:

1. From 2WD parking area, follow the rocky, dirt road for 0.5 miles to reach the Hessie trailhead.
2. Cross the bridge, then stay on the main dirt road up.
3. Follow signs for Woodland Lake; trail junctions are well marked.
4. Around 5 miles, reach open view of Woodland Lake to your left.
5. Continue up trail for another quarter-mile to visit Skyscraper Reservoir before returning the way you came.

FIND THE TRAILHEAD

On summer weekends, take the free Hessie Trailhead shuttle from the RTD Park-n-Ride in Nederland. Otherwise, head out early for a good shot at the limited parking. From Nederland, drive southwest out of town on CO 119, then turn right on CO 130/Eldora Road. In 4.5 miles, arrive at Hessie 2WD Trailhead at the junction with Fourth of July Road. Park on the road in designated areas only.

VERY NICE BREWING COMPANY

This cozy, family-owned establishment sports a 3.5-barrel brewing system and a lineup of six flagship beers as well as a rotating menu of seasonals and premium beers. Locals are invited to bring in their own ingredients to contribute to small-batch experimental brews as well. With frequent live music and a family-friendly atmosphere, it's hard not to fall in love with this quaint little neighborhood watering hole.

CONTACT INFORMATION
U.S. Forest Service,
Boulder Ranger District,
2140 Yarmouth Ave
Boulder, CO 80301;
303-541-2500

BREWERY/RESTAURANT
Very Nice Brewing Company
20 Lakeview Drive, Suite 112,
Nederland, CO 80466
303-258-3770
Miles from trailhead: 5.3

NORTH BOULDER

A STEEP, SHORT TREK NORTHWEST OF TOWN

▷⋯ STARTING POINT	⋯✕ DESTINATION
2ND STREET AND DAKOTA BOULEVARD	**HOGBACK RIDGE**
🍺 BEER	🔲 HIKE TYPE
THAI STYLE WHITE IPA	**MODERATE-STRENUOUS**
$ FEES	📅 SEASON
NONE	**YEAR-ROUND**
⛰ MAP REFERENCE	🐾 DOG FRIENDLY
FOOTHILLS AREA TRAIL MAP AT BOULDERCOLORADO.GOV	**NO**
🕐 DURATION	↦ LENGTH
1.5 HOURS	**2.4 MILES**
↑↓ LOW POINT / HIGH POINT	〰 ELEVATION GAIN
5,670 FEET / 6,417 FEET	**700 FEET**

WHITE IPA

THAI STYLE
WHITE IPA

SLIGHTLY HAZY
GOLDEN

CITRUS, LEMONGRASS,
BELGIAN YEAST

LIGHT-BODIED;
PEPPER, CORIANDER

BITTERNESS
IBU: 33

SWEETNESS

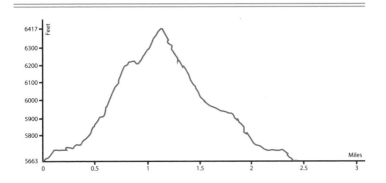

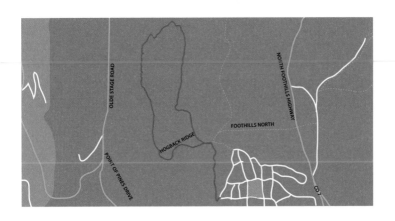

DESCRIPTION OF THE ROUTE

Escape the Boulder trail crowds by venturing onto this steep, hiker-only loop northwest of town, then wash it down with a "Thai IPA" brewed with seven Asian spices.

Boulder frequently appears on "Best Trail Town" lists—and for obvious reasons, with more than 200 miles of trails within easy range of downtown. Because Boulder's population has boomed in recent years, many of the most popular trails can get quite crowded, especially on summer weekends.

The solution? Venture to the lesser-known singletrack systems northwest of town on the outskirts of North Boulder, or "NoBo" as it's called colloquially. Because neither dogs nor mountain bikers are permitted on this loop, it's a great choice for hikers desiring a bit of solitude outside a city otherwise bustling with fellow outdoors enthusiasts.

"Hogback" is an actual geological term for a long, narrow ridge created over time by erosion. Colorado is chock-full of trails called "Hogback" that traverse such ridgelines, and this one is no different. It's a relatively steep climb up, but absolutely worth the short trek for expansive views of the eastern plains, sparkling reservoirs, neighboring foothills, the Boulder Flatirons, and Mount Sanitas.

Wear a hat and ample sunscreen, as there's hardly any shade on this trail other than the brief respite offered by some scrubby pines near the top of the hogback itself. Otherwise, the trail is mostly desert grasses and rock or log stairs. Admire the beautifully designed rock staircases you encounter along the way.

The Hogback Loop is a lollipop loop that begins with a section of the Foothills Trail. After a quarter mile on the Foothills Trail, veer left and head uphill to begin the Hogback loop. Stay to the left when the trail splits to continue on a clockwise circuit of the loop. The trail continues by descending the hogback and closes the loop just before rejoining Foothills Trail.

Do keep an eye out for rattlesnakes, as they enjoy sunning themselves on the rocks along this trail. Other wildlife sightings are common as well, including mule deer and coyotes. After soaking up the views at the top, descend on the north side to complete a clockwise lollipop loop before heading to Upslope Brewing's NoBo taproom less than a mile away on Lee Hill Drive.

TURN BY TURN DIRECTIONS:

1. From the trailhead, follow the clear uphill trail to your right.
2. At 0.3 miles, go left at the junction, then arrive at the well-signed Hogback Loop, which can be hiked in either direction.
3. If you go left (clockwise), reach the high point at 1.1 miles. Scramble up to your right to reach the very top before rejoining the trail.
4. Continue on the loop, descending north.
5. At 2.1, rejoin the original trail by going left at the first junction and right at the second junction.

FIND THE TRAILHEAD

Go west on Lee Hill Drive out of North Boulder. Make a right on 5th Street, then a left on Dakota Boulevard. The small trailhead is at the corner where Dakota meets 2nd Street. Park on the street.

UPSLOPE BREWING COMPANY - LEE HILL

Upslope is named for the meteorological phenomenon responsible for many a snowstorm in Colorado's Front Range. The name is an ode to Upslope's founders' love for snow, mountains, and adventure. When the team of three first opened Upslope in 2008, their mission was to "make a better beer rooted in the outdoor lifestyle." They have five core styles, with a number of popular seasonals ranging from a pumpkin ale to a decadent Christmas ale to this Thai-style White IPA.

CONTACT INFORMATION
Open Space and Mountain Parks (OSMP),
3198 N. Broadway,
Boulder, CO 80304; 303-441-3440

BREWERY/RESTAURANT
Upslope Brewing Company - Lee Hill
1501 Lee Hill Drive,
Boulder, CO 80304
303-449-2911
Miles from trailhead: 0.8

BOULDER BEAR PEAK

ASCEND A BELOVED LOCAL PEAK VIA THE FERN CANYON TRAIL

▷⋯ STARTING POINT	⋯✗ DESTINATION
NCAR TRAILHEAD	**BEAR PEAK**
🍺 BEER	🉐 HIKE TYPE
ANNAPURNA AMBER	**VERY STRENUOUS**
$ FEES	📅 SEASON
NONE	**MARCH TO NOVEMBER**
🗺 MAP REFERENCE	🐾 DOG FRIENDLY
OSMP TRAIL MAP AT BOULDERCOLORADO.GOV	**YES (LEASH REQUIRED)**
🕐 DURATION	↦ LENGTH
4-6 HOURS	**7.5 MILES**
↑↓ LOW POINT / HIGH POINT	〜 ELEVATION GAIN
6,109 FEET / 8,461 FEET	**2,687 FEET**

ALCOHOL 4.7% CONTENT

AMBER

 BROWN, REDDISH TINT

 BISCUIT, CARAMEL

 TOFFEE, BARLEY MALT

BITTERNESS
IBU: UNLISTED

SWEETNESS

SOUTHERN SUN
Pub & Brewery

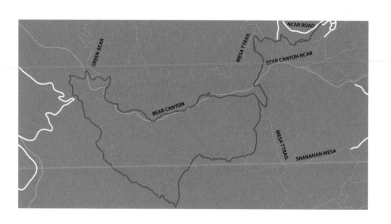

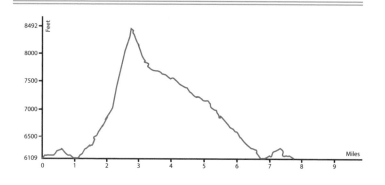

DESCRIPTION OF THE ROUTE

Go for this "bear" of a climb and scramble up a famous local peak before visiting one of Boulder's original brewpubs.

The town of Boulder is framed on the west by a number of prominent mountains and ridgelines visible from most parts of downtown. These include the infamous Flatirons on the front flanks of Green Mountain, as well as Bear Peak and South Boulder Peak. With many trails weaving over their shoulders, all offer spectacular, challenging hiking and a bit of scrambling. From their summits, you'll enjoy spectacular views of Boulder to one side and the vast Indian Peaks Wilderness to the other.

If you're in the mood for a slightly shorter hike, you can do this summit of Bear Peak as an out-and-back in a little over five miles. However, the longer descent through Bear Canyon provides a lovely amble through the trees and over a number of creek crossings—and it's a more forgiving route on the knees and quads than coming back down the much steeper Fern Canyon Trail you'll ascend on your way up.

From the National Center for Atmospheric Research (NCAR) trailhead, the initial approach on the Mesa Trail passes through dry, dusty desert environs. It can get quite hot in the middle of the day during summer. But it won't be long before you get to duck into the trees and seek respite among the pines.

Once you turn onto the lush Fern Canyon Trail, be prepared for a staircase of a workout for about one mile until you reach the high, open saddle just beneath the 8,459-foot summit of Bear Peak, the second highest mountain in Boulder. From there, you can see the adjacent South Boulder Peak (8,549 feet) to the west, as well as the larger mountains along the Continental Divide on the horizon line. To the east, you get a bird's-eye view of Boulder and rolling plains that lead to Denver and beyond.

From here, you'll be able to see the Bear Peak West Ridge Trail snaking down the ridge to your right. A trail sign below the summit marks the turnoff for this route to Bear Canyon. When you're ready, and assuming you want to complete the full loop, continue in this direction for a descent that's initially steep but quickly mellows to a more flowing, roundabout hike through wildflower-studded meadows and mixed forest.

Before you go, check www.OSMP.org for current dog/leash regulations and trail conditions; portions of this route are subject to frequent closures. This route can generally be done year-round, though if you plan to hike it in the winter, traction devices (spikes) are highly recommended.

TURN BY TURN DIRECTIONS:

1. Stay left at the first two trail junctions at 0.6 miles and 0.7.
2. At 1.4 miles, make a left, followed by a right onto Fern Canyon at 1.5.
3. At 1.9, go right and up.
4. At 2.7, scramble up to reach the Bear Peak summit.
5. Follow signs for Bear Peak West Ridge (the trail winding down the ridge below to the northwest).
6. At 4.6, stay straight to stay on Bear Canyon Trail.
7. At 6.75, go left and up on the Mesa Trail, following signs for NCAR, then stay right at next three trail intersections.

FIND THE TRAILHEAD

From S. Broadway (Highway 93), drive west out of South Boulder on Table Mesa Drive. This turns into NCAR Road before arriving at the large, well-signed trailhead at the end of the road.

SOUTHERN SUN PUB & BREWERY

Southern Sun Pub & Brewery is one of three local branches of the Mountain Sun, a bohemian-themed neighborhood brewpub that first opened its doors back in 1993. Its south Boulder locale makes it a perfect stop after a day of hiking in nearby Chautauqua Park, home to Boulder's iconic Flatirons and many other beloved local hikes including this one. Grab a seat on the outdoor patio or a booth by the window for a great sunset view over the mountains.

CONTACT INFORMATION
Boulder Open Space and Mountain Parks,
66 S. Cherryvale Road,
Boulder, CO 80303;
303-441-3440

BREWERY/RESTAURANT
Southern Sun Pub & Brewery
627 S. Broadway,
Boulder, CO 80305
303-543-0886
Miles from trailhead: 2.6

BOULDER BOBOLINK

A GENTLE, URBAN STROLL THROUGH A RIPARIAN CORRIDOR

▷⋯ STARTING POINT	⋯✗ DESTINATION
BOBOLINK TRAILHEAD	**SOUTH BOULDER CREEK**
🍺 BEER	HIKE TYPE
PONDEROSA PORTER	**URBAN WALK**
$ FEES	SEASON
NONE	**YEAR-ROUND**
⌂ MAP REFERENCE	🐾 DOG FRIENDLY
BOBOLINK-CHERRYVALE MAP AT BOULDERCOLORADO.GOV	**YES (LEASH REQUIRED)**
🕐 DURATION	↦ LENGTH
30-45 MINUTES	**1.25 MILES**
↑↓ LOW POINT / HIGH POINT	〰 ELEVATION GAIN
5,295 FEET / 5,341 FEET	**MINIMAL**

PORTER

 MIDNIGHT BLACK, RED HUE

COFFEE, BUTTERSCOTCH, PINE

VANILLA, RYE MALT

BITTERNESS
IBU: 40

SWEETNESS

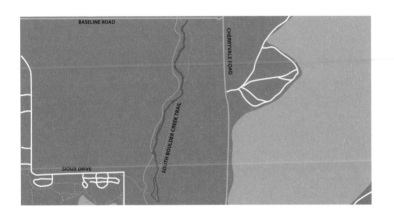

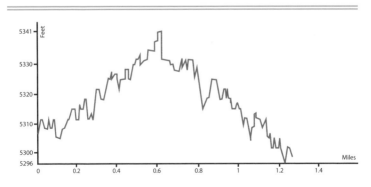

DESCRIPTION OF THE ROUTE

A gentle stroll through a riparian corridor and birdwatcher's paradise, followed by a visit to the nature-inspired Wild Woods Brewery.

The Bobolink is a small blackbird native to North America, whose unique warble Henry David Thoreau once characterized by saying, "It is as if he touched his harp within a vase of liquid melody, and when he lifted it out the notes fell like bubbles from the trembling strings. Methinks they are the most liquidly sweet and melodious sounds I ever heard."

This mellow, wheelchair-accessible path meanders along South Boulder Creek on a flat, wide dirt path.Interpretive signs along the way offer information about the bobolink and other nesting fowl that can be observed in the neighboring meadow to the east, most prevalently in late spring and through the summer. Also keep an eye out along your way for white-tailed deer, mule deer, red-tailed hawks, coyotes, and foxes.

As you wander along the wooden fence lined with colorful cottonwoods, enjoy peeks of the stately Flatirons through the trees. Several designated creek access points along the way let you wander down to the water and, if you feel so inclined, dip your toes in. A rock beach adorned with cairns makes a nice picnic spot.

At 0.6 miles, your route will intersect with a multi-use trail, the South Boulder Creek Trail, that roughly parallels the path you've been walking on. If you'd like to extend your walk, you can hop on that and continue south along the river for another mile (one way) to South Boulder Road. The South Boulder Creek Trail continues for another two miles beyond that—all the way to Marshall Road—though note that dogs are not permitted on this stretch.

TURN BY TURN DIRECTIONS:

1. Follow the dirt trail along the creek.
2. At 0.6 miles, turn around and head back—or extend your stroll on the South Boulder Creek Trail for 1-3 more miles (one way).

FIND THE TRAILHEAD

Head east out of downtown Boulder on Arapahoe Avenue, then turn right onto Foothills Parkway. In 1 mile, make a left onto Baseline Road. The Bobolink trailhead will be on your right in 1 mile, across from the intersection with Gapter Road.

WILD WOODS BREWERY

Wild Woods Brewery opened in the fall of 2012 as a two-barrel nanobrewery started by longtime home brewers, hikers, and camping enthusiasts Jake and Erin Evans. Dreamt up around a campfire in Crested Butte, Colorado, their mission was to create handcrafted beer inspired by the outdoors. They've since expanded to a seven-barrel brewing system, and their flagship brews incorporate inspiration from local wildflowers, berries, juniper, and ponderosa pine trees and other wild ingredients.

CONTACT INFORMATION
Boulder Open Space and Mountain Parks,
66 S. Cherryvale Road,
Boulder, CO 80303;
303-441-3440

BREWERY/RESTAURANT
Wild Woods Brewery
5460 Conestoga Court,
Boulder, CO 80301
303-484-1465
Miles from trailhead: 2.1

ESTES PARK

A DAZZLING ALPINE LAKE IN ROCKY MOUNTAIN NATIONAL PARK

▷⋯ STARTING POINT	⋯✕ DESTINATION
LONGS PEAK TRAILHEAD	**CHASM LAKE**
⬗ BEER	▦ HIKE TYPE
SMOKY BRUNETTE SMOKED BROWN ALE	**STRENUOUS**
$ FEES	📅 SEASON
NONE	**JUNE TO OCTOBER**
△ MAP REFERENCE	🐾 DOG FRIENDLY
TRAILS ILLUSTRATED 200: ROCKY MOUNTAIN NATIONAL PARK	**NO**
🕐 DURATION	↦ LENGTH
5-7 HOURS	**8.4 MILES**
↑↓ LOW POINT / HIGH POINT	〜 ELEVATION GAIN
9,426 FEET / 11,847 FEET	**2,500 FEET**

BROWN ALE

 AMBER

 BREAD, TOFFEE, GERMAN MALT

 INTENSELY SMOKY, NUTTY

BITTERNESS	SWEETNESS
IBU: 26	

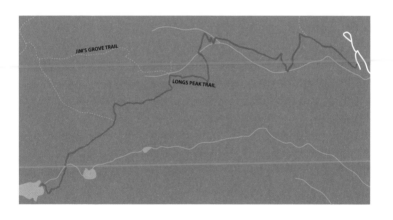

JIM'S GROVE TRAIL

LONGS PEAK TRAIL

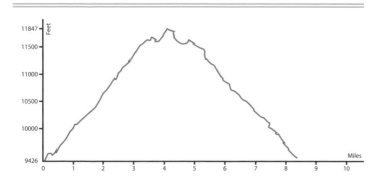

DESCRIPTION OF THE ROUTE

This dazzling alpine lake sits at the foot of one of Colorado's most iconic, awe-inspiring mountain faces.

The crown jewel of Rocky Mountain National Park, Longs Peak is its tallest peak and only "fourteener," standing proud at 14,259 feet. It is featured on the Colorado state quarter, and its dramatic, sheer east face is known as "The Diamond" for reasons that will become obvious during this hike. Chasm Lake lies at the foot of the Diamond—a gorgeous alpine lake burrowed into the belly of a dramatic cirque, carved out long ago by glaciers. The one remaining sliver of a glacier on Longs Peak, Mills Glacier, sits just above the lake.

Find your trail behind the ranger station. You'll begin by hiking through subalpine lodgepole pine forest before emerging above tree line. Listen for the squeaks of pikas and chirps of marmots darting amidst the rocks and wildflower-pocked tundra. You'll pass the turnoff for the Boulder Field, where you'll part ways from hikers bound for the summit of Longs. Beyond there, admire the face of the Diamond as you traverse the alpine hillside toward it, passing Peacock Pool below to your left. Depending on the time of year you go, a waterfall often spills down between here and the rocky bluffs above.

There's a small scramble in the last tenth of a mile to reach Chasm Lake, as the "trail" snakes up and through a notch in the rock. Follow cairns to find your route. Enjoy the deep emerald hue of the lake with its beautiful reflection of the surrounding jagged peaks, snowfields, and rocky cliffs. Plenty of flat boulders around the lake's edge let you relax in the sun. If you've brought binoculars, it's possible to watch climbers ascending Longs Peak, which looms over the lake like a sentry.

Because this is also the trailhead for Longs Peak, the parking lot can fill up fast, especially on summer weekends. Plan to arrive early. Above tree line, high wind, harmful UV rays, and lightning can all be threats—so, wear sunscreen, bring layers, and be prepared to turn back early if thunderstorms begin to develop.

TURN BY TURN DIRECTIONS:

1. At 0.5 mile, go straight at the well-signed junction, following signs for Chasm Lake and Longs Peak.
2. At 2.5, go left at the well-signed junction.
3. At 3.2, stay straight/left at the junction with the Boulder Field.
4. At 4, scramble up the steep gully, following cairns.
5. In a tenth of a mile, reach the lake.

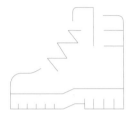

FIND THE TRAILHEAD

From Estes Park, head south on either Mary's Lake Road or South St. Vrain Avenue (Highway 7). About 5.5 miles farther south of where these two roads intersect, turn right onto Longs Peak Road. Arrive at the trailhead in 1 mile.

ROCK CUT BREWING COMPANY

Rock Cut Brewing Company offers a cozy, après-adventure spot to quench your thirst and rest your weary legs after hiking in nearby Rocky Mountain National Park. Located at the foot of Prospect Mountain with an outdoor patio overlooking Lumpy Ridge, this family-friendly taproom pours everything from crisp IPAs to barrel-aged imperial stouts to a number of German-style brews, like this smoky brown ale.

CONTACT INFORMATION
National Park Service,
Rocky Mountain National Park,
1000 U.S. Highway 36,
Estes Park, CO 80517;
970-586-1206

BREWERY/RESTAURANT
Rock Cut Brewing Company
390 W. Riverside Drive,
Estes Park, CO 80517
970-586-7300
Miles from trailhead: 10.3

FORT COLLINS
HORSETOOTH ROCK

TAKE THE HORSETOOTH ROCK TRAIL TO A FUN SUMMIT SCRAMBLE

▷⋯ STARTING POINT	⋯✕ DESTINATION
HORSETOOTH MOUNTAIN TRAILHEAD	**HORSETOOTH FALLS AND HORSETOOTH ROCK**
BEER	HIKE TYPE
ZWEI PILS BAVARIAN-STYLE PILSNER	**STRENUOUS**
$ FEES	SEASON
$9	**MARCH TO NOVEMBER**
MAP REFERENCE	DOG FRIENDLY
HORSETOOTH MOUNTAIN OPEN SPACE ON LARIMER.ORG	**YES (LEASH REQUIRED)**
DURATION	LENGTH
3-4 HOURS	**6.1 MILES**
↑↓ LOW POINT / HIGH POINT	∿ ELEVATION GAIN
5,745 FEET / 7,255 FEET	**1,585 FEET**

ALCOHOL 5.1% CONTENT

PILSNER

 CLEAR, PALE STRAW

FLORAL, HONEY AND BISCUITS

CRISP, GRASSY, HERBAL HOPS

BITTERNESS
IBU: 24

SWEETNESS

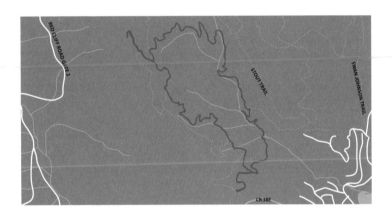

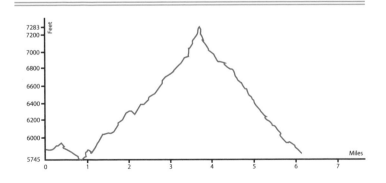

DESCRIPTION OF THE ROUTE

No Fort Collins trip is complete without a trip to Horsetooth Mountain Open Space; this loop encompasses a waterfall, fun summit scramble, and an authentic German-style pilsner back in town.

Horsetooth Mountain—so named for its resemblance to giant, literal horse's teeth—is a landmark on the western horizon of Fort Collins. It's even a part of the official City of Fort Collins logo. According to Arapahoe legend, the rock is what remains of the heart of giant who once presided over the area before Chief Maunamoku killed him, cleaving the giant's chest into the rock formation we now see today.

Visible from many parts of town, it is an icon that beckons to be climbed.

Fortunately, nearly 30 miles of trails have been developed in the public open space that sits at the foot of the mountain. Often referred to as Horsetooth Rock by locals, Horsetooth Mountain is not only a doable (and justly popular) day hike, there are multiple route options for reaching its summit.

The most direct, and thus most crowded, hiking route up is the Horsetooth Rock Trail. This loop utilizes that trail on the way down, but on the way up, detours onto the Horsetooth Falls Trail—also a nice option for a shorter, easier hike if you don't want to venture all the way to the top. The falls are modest (and may be gone altogether during some times of the year such as autumn), but are a lovely spot nonetheless to enjoy a snack, tucked away in a quiet ravine in the woods. Reaching the summit via this trail permits a more forgiving grade on the climb up.

The dense ponderosa forest is home to mule deer, fox, black bear, and many species of birds, so wildlife enthusiasts will love this hike. As you near the top of Horsetooth Rock, the hike turns into a scramble, requiring a few Class 2 moves to make it to either the north or south summit rocks. From the top, the 360-degree views are breathtaking. Take your time up here, savoring glimpses of the nearby Mummy Range and much-adored 14er Longs Peak.

TURN BY TURN DIRECTIONS:

1. Start up Horsetooth Rock Trail to the right and continue following signs for Horsetooth Falls.
2. At 1 mile, go straight/left for a brief out-and-back to a viewpoint of the falls before continuing on up Spring Creek Trail.
3. At 2.1, go left onto Wathen.
4. At 3.3, go right onto Horsetooth Rock Trail and scramble to summit at 3.7 miles.
5. Head back down to junction with Wathen and go right this time.
6. Follow signs saying "Attention: Foot Travel Only" to stay on Horsetooth Rock Trail on your way down.
7. At 5.2, go right on Soderberg, then right again a half-mile later at the original junction with the Horsetooth Falls Trail.

FIND THE TRAILHEAD

South of downtown Fort Collins, take County Road 38E west to Red Cliff Road. The trailhead parking area for Horsetooth Mountain Open Space is to your right. Ample parking exists, but the lot can still fill up on busy days, so plan to arrive early to ensure a parking spot.

ZWEI BREWING

This German-inspired brewery in Fort Collins crafts authentic German, Belgian, and Czech lagers, as well as some small-batch American-style creations such as IPAs, red ales, and brown ales. Their four flagship beers are a southern German-style pilsner, Munich-style golden helles, Munich-style amber dunkel, and a Bavarian-style Weissbier wheat. A sunny outdoor patio with lawn games, a ping-pong table and rotating food trucks on site round out this lively taproom experience.

CONTACT INFORMATION
Larimer County Open Space,
200 W Oak Street,
Fort Collins, CO 80521;
970-498-7000

BREWERY/RESTAURANT
Zwei Brewing
4612 S Mason Street, Suite 120,
Fort Collins, CO 80525
970-223-2482
Miles from trailhead: 8.9

FORT COLLINS RESERVOIR RIDGE

A QUIET, DESERT-ENVIRONS HIKE ON THE NORTH LOOP TRAIL

▷⋯ STARTING POINT	⋯✗ DESTINATION
MICHAUD TRAILHEAD	**NORTH LOOP TRAIL**
🍺 BEER	🔁 HIKE TYPE
JONAS PORTER	**MODERATE**
$ FEES	📅 SEASON
NONE	**YEAR-ROUND**
🗺 MAP REFERENCE	🐾 DOG FRIENDLY
RESERVOIR RIDGE TRAIL MAP AT FCGOV.COM	**YES (LEASH REQUIRED)**
🕐 DURATION	↦ LENGTH
2 HOURS	**3.75 MILES**
↑↓ LOW POINT / HIGH POINT	〰 ELEVATION GAIN
5,167 FEET / 5,643 FEET	**460 FEET**

 PORTER

 OPAQUE BLACK

EARTH, BLACK LICORICE

CREAMY; COFFEE, DARK MALTS

BITTERNESS	SWEETNESS
IBU: 39	

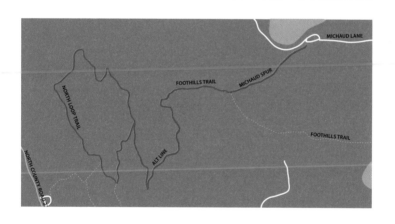

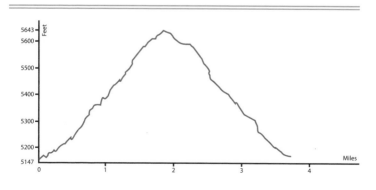

DESCRIPTION OF THE ROUTE

A quiet, desert-environs alternative to more crowded trails, this loop is a great place to spot mule deer, rattlesnakes, and wildflowers in the summer.

Fort Collins' most popular trails sit in Horsetooth Mountain Open Space at the south end of Horsetooth Reservoir. To beat the crowds (and user fees at Horsetooth), venture to a quieter trail system at the north end of the reservoir, the aptly named Reservoir Ridge area. Here, desert environs prevail, with many opportunities to catch glimpses of mule deer meandering in the tall grass.

Several trailhead options exist for accessing the trails on Reservoir Ridge. (In fact, the 9.6-mile Foothills Trail connects this area to a number of other local trails and foothills along the reservoir, including the Pineridge and Maxwell natural areas.) This loop begins from the trailhead at the west end of Michaud Lane, right next to Claymore Lake, a private but nevertheless picturesque little body of water to admire on your hike up through the grasslands.

The climb up these foothills is moderate—never too steep, but quite rocky for much of the way. Watch your footing for rocks and other obstacles, including the possibility of rattlesnakes. As you initiate a clockwise loop on the North Loop Trail, glance back over your left shoulder for a view of the north end of Horsetooth Reservoir. A few interpretive signs along the next stretch of trail also offer information about the area.

If you can believe it, 70 percent of Colorado's craft beer is produced in Fort Collins (thanks in large part to the biggest three—Odell, New Belgium, and Fort Collins Brewery), so options abound for craft beers afterward! Some 20-plus breweries call Fort Collins home—but Equinox Brewing, opened in 2010, tends to stay especially bustling in the evenings. One of their more popular brews, the Jonas Porter, makes an especially roasty dessert for your efforts—but since Equinox has as many as 20 rotating taps at any given time, you can't go wrong with a sampler flight either!

TURN BY TURN DIRECTIONS:

1. After a third of a mile, go right at the signed junction.
2. At 1.25 miles, go left at an unmarked junction to initiate a clockwise loop.
3. At 1.5, go right, then right again, to follow the North Loop Trail.
4. At 2.5, turn left to return to the trailhead.

FIND THE TRAILHEAD

From downtown Fort Collins, drive west to reach N Overland Trail (road). Turn right. Follow this until you can make a left onto Michaud Lane. The trailhead is at the end of this road.

EQUINOX BREWING

Fort Collins is a hotbed for craft beer—and Equinox lives up to its hometown's reputation, with rotating taps that have featured more than 100 different beers since its opening in 2010. Centrally located downtown, its lively taproom and outdoor beer garden are always bustling with activity, including shuffleboard, food trucks, and live music on the weekends.

CONTACT INFORMATION
City of Fort Collins Parks,
413 South Bryan Ave,
Fort Collins, CO 80521;
970-221-6660

BREWERY/RESTAURANT
Equinox Brewery
133 Remington Street,
Fort Collins, CO 80524
970-484-1368
Miles from trailhead: 5.7

LOUISVILLE LAFAYETTE

A SHORT AND SWEET URBAN HIKE ALONG STEARNS LAKE

▷⋯ STARTING POINT	⋯✕ DESTINATION
STEARNS LAKE TRAILHEAD	**NORTH SIDE OF STEARNS LAKE**
🍺 BEER	🔁 HIKE TYPE
TREACHERY GOLDEN STRONG ALE	**URBAN WALK**
$ FEES	📅 SEASON
NONE	**YEAR-ROUND**
⛰ MAP REFERENCE	🐾 DOG FRIENDLY
CAROLYN HOLMBERG PRESERVE TRAIL MAP AT BOULDERCOUNTY.ORG	**YES (LEASH REQUIRED)**
🕐 DURATION	↦ LENGTH
45 MINUTES	**1.2 MILES**
↕ LOW POINT / HIGH POINT	〰 ELEVATION GAIN
5,292 FEET / 5,312 FEET	**MINIMAL**

ALCOHOL 8% CONTENT

ALE

HAZY GOLDEN

CORIANDER, CLOVES, BUBBLEGUM ESTERS

ORANGE ZEST, BELGIAN GRAINS

BITTERNESS
IBU: 25

SWEETNESS

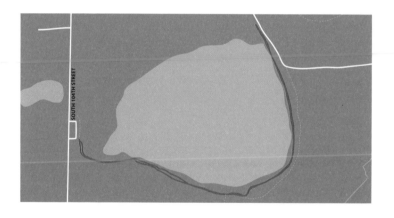

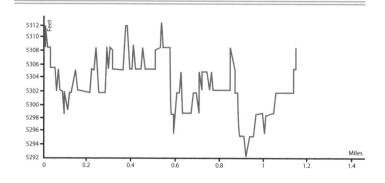

DESCRIPTION OF THE ROUTE

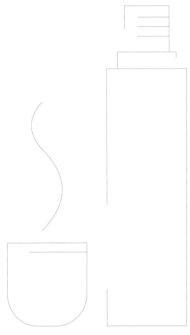

Bordering Stearns Lake, this urban hike is short and sweet, just like the Belgian-style strong pale ale that accompanies it.

If you're in the mood for a short but thoroughly relaxing stroll, this little gem northwest of Denver is the ticket. For being in the heart of the suburbs, this is a remarkably quiet, often secluded park. If there's any hint of a breeze in the air, the sound of Stearns Lake's water lapping at the edges of its shore drowns out any of the distant murmur of civilization.

The trail's namesake, Mary E. Miller, was the wife of a gold prospector, Lafayette Miller, who arrived in the area in the early 1870s. After her husband died suddenly in 1878, Mary raised their six children on her own, began coal mining on her own property, became the first female bank president in the world, and founded the town of Lafayette.

This short-and-sweet out-and-back hits the highlight of Stearns Lake. You can continue northeast for another mile, and even longer in either direction if you link up with neighboring trails on either end—so it's possible to extend this hike for much longer than what's suggested here. (Another half-mile to the northeast, for example, there is a sizable prairie-dog habitat.)

From the lake's northern shore, glance back across the bright blue water for glimpses of Boulder's iconic Flatirons and the greater Indian Peaks range behind them. It's also possible to fish in the lake, as it's regularly stocked with bluegill, channel catfish, tiger muskie, and largemouth bass. And, birdwatching enthusiasts rejoice; this is a popular haunt for red-tailed hawks, eagles, sparrows, geese, and numerous other waterfowl.

TURN BY TURN DIRECTIONS:

1. Follow the flat, gravel path around right side of lake.
2. A quarter-mile in, stay left/straight to continue on Mary Miller Trail.
3. Veer left of the main gravel trail to walk on top of the dam for better views.
4. Reach northeast corner of Stearns Lake at 0.6 miles. Turn around here.

FIND THE TRAILHEAD

From downtown Louisville, drive south on Courtesy Road/96th Street. Make a left on W. Dillon Road, then a right onto S. 104th Street. In less than a mile, the Stearns Lake parking lot will be on your left.

12DEGREE BREWING

Like many microbreweries in Colorado, 12Degree began as a homebrewer's pipe dream. Jon Howland was enamored with the saisons, strong ales, quadrupels, and sour beers he encountered while backpacking in Belgium. He opened up 12Degree in 2013 in—literally!—a former "temperance house," an alcohol-abstinence bar. One of its flagship brews, the Treachery Golden Strong Ale, snagged 2016 honors at the Great American Beer Festival and the World Beer Cup. Just pace yourself; even with a modest 10-ounce pour, it hides its 8-percent ABV well. 12Degree's flatbread pizzas, cheese plates, and chocolates complement the beer beautifully.

CONTACT INFORMATION
Boulder County Parks and Open Space,
5201 Saint Vrain Road,
Longmont, CO 80503;
303-678-6200

BREWERY/RESTAURANT
12Degree Brewing
820 Main Street,
Louisville, CO 80027
720-638-1623
Miles from trailhead: 3.4

DENVER
ROCKY MOUNTAIN ARSENAL

A QUIET AMBLE THROUGH AN URBAN NATIONAL WILDLIFE REFUGE

▷⋯ STARTING POINT	⋯✕ DESTINATION
ROCKY MOUNTAIN ARSENAL CONTACT STATION	**LAKE MARY AND LAKE LADORA**
🍺 BEER	🔄 HIKE TYPE
JUICY BANGER IPA	**URBAN WALK**
$ FEES	📅 SEASON
NONE	**YEAR-ROUND**
⛰ MAP REFERENCE	🐾 DOG FRIENDLY
ROCKY MOUNTAIN ARSENAL MAP AT FWS.GOV	**NO**
🕐 DURATION	↦ LENGTH
1-1.5 HOURS	**3 MILES**
↕ LOW POINT / HIGH POINT	〰 ELEVATION GAIN
5,210 FEET / 5,253 FEET	**MINIMAL**

ALCOHOL 7.4% CONTENT

IPA

STATION 26 BREWING co
JUICY BANGER IPA

 GOLDEN, SPARKLING

 APRICOT, GRAPEFRUIT

 SPRITZY PINEAPPLE, PINEY HOPS

BITTERNESS
IBU: 100+

SWEETNESS

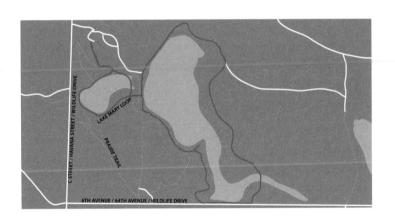

C STREET / HAVANA STREET / WILDLIFE DRIVE

LAKE MARY LOOP

PRAIRIE TRAIL

6TH AVENUE / 64TH AVENUE / WILDLIFE DRIVE

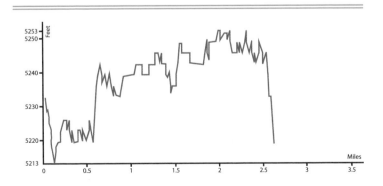

DESCRIPTION OF THE ROUTE

Enjoy a quiet stroll through a former military arsenal site that's been converted into one of the country's largest urban national wildlife refuges.

Located on the northeast fringe of Denver and managed by the U.S. Fish and Wildlife Service, Rocky Mountain Arsenal Wildlife Refuge boasts a whopping 15,000 acres of wooded areas, wetlands, and native short grass prairie. Some 330 species of wildlife call the Arsenal home, including bison, bald eagles, red-tailed hawks, migratory song birds, owls, deer, coyotes, and prairie dogs. Ten miles of trails are open year-round (from sunrise to sunset) for hiking, snowshoeing, and cross-country skiing.

Once upon a time, the Refuge's 15,000 acres were home to large herds of bison, Plains Indians, and, eventually, early American settlers and farmers. After the attack on Pearl Harbor in World War II, the area was converted into a chemical weapons manufacturing facility, then later used for weapons production during the Cold War. In the 1980s, the government initiated an environmental cleanup of the area and, a decade later, turned it over to the U.S. Fish and Wildlife Service to manage as a national wildlife refuge.

It's possible to take a self-guided, 11-mile driving tour through the Arsenal and hit up many of the highlights, but this short, easy loop trail lets you do a shorter tour on foot. From the outset of your hike, you may be able to glimpse bison grazing in the fields across the street to your right. Around both Lake Mary and Lake Ladora, keep an eye out for nesting waterfowl, bald eagles, prairie dogs, and other critters. As you make your way around Lake Ladora, you'll enjoy a nice view of the Denver skyline silhouetted against the snowy Front Range peaks to the west. Several interpretive signs along the way offer information on local fauna and flora.

There's very little shade on this trail other than the occasional smattering of impressively tall cottonwoods along the lakeshore, so if it's sunny out, be sure to wear a hat and sunscreen.

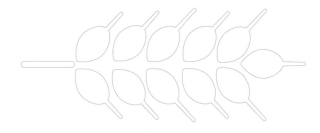

TURN BY TURN DIRECTIONS:

1. Start by following signs for Lake Mary Loop and stay left at first junction.
2. At 0.15 miles, go right for a counterclockwise loop of Lake Mary.
3. At 0.55, stay right to climb the log staircase up to Lake Ladora.
4. Follow the trail around Lake Ladora counterclockwise. At 1.2 miles, do not cross the paved road. Parallel the left side of it on the gravel shoulder for a quarter-mile, then go left again onto Lake Ladora Loop Trail.
5. At 2.5, return down stairs you came up, then go right to cross floating bridge and complete your loop.

FIND THE TRAILHEAD

This hike begins at the contact station in Rocky Mountain Arsenal Wildlife Refuge. The Arsenal Visitor Center is located at 6550 Gateway Road in Commerce City (northeast of Downtown Denver and Station 26 Brewing). Get a map to locate the contact station and, from the visitor center, follow signs for Lakes/Trails until you reach the parking area just north of Lake Mary.

STATION 26 BREWING

Station 26 Brewing is located inside an old firehouse, with a beautiful outdoor patio to boot. They brew all kinds of beer, ranging from a tangerine cream ale to a silky porter to a lemondrop wheat, but their IPAs—both those on tap at the firehouse and their ever-evolving, single-hop series (available in cans)—are especially off the charts. Depending on the day of the week, various local food trucks park onsite to offer brewery patrons everything from pizza to tacos to barbecue.

CONTACT INFORMATION
U.S. Fish and Wildlife Service,
Rocky Mountain Arsenal Visitor Center,
6550 Gateway Road,
Commerce City, CO 80022;
303-289-0930

BREWERY/RESTAURANT
Station 26 Brewing
7045 E 38th Avenue,
Denver, Colorado 80207
303-333-1825
Miles from trailhead: 6

DENVER LAKEWOOD

EASY EXPLORATIONS IN A NEIGHBORHOOD PARK

▷··· STARTING POINT	···✕ DESTINATION
YARROW STREET AND W. KENTUCKY AVENUE	**KOUNTZE LAKE**
🍺 BEER	🎋 HIKE TYPE
OLD 121 HONEY BROWN ALE	**URBAN WALK**
$ FEES	📅 SEASON
NONE	**YEAR-ROUND**
⛰ MAP REFERENCE	🐾 DOG FRIENDLY
BELMAR PARK AT MAPS.GOOGLE.COM	**YES**
🕐 DURATION	↦ LENGTH
45 MINUTES	**1.6 MILES**
↑↓ LOW POINT / HIGH POINT	〰 ELEVATION GAIN
5,482 FEET / 5,554 FEET	**MINIMAL**

ALCOHOL 6% CONTENT

BROWN ALE

 CLEAR CHESTNUT BROWN

 MILD HONEY, COFFEE

 EARTHY, LIGHT-BODIED

OLD·121

SOUTH WADSWORTH BLVD

BITTERNESS
IBU: 25

SWEETNESS

5
4
3
2
1

5
4
3
2
1

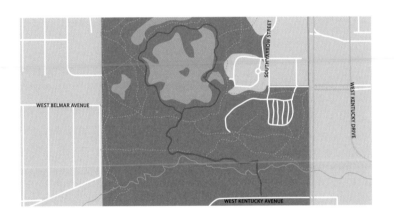

WEST BELMAR AVENUE

SOUTH YARROW STREET

WEST KENTUCKY DRIVE

WEST KENTUCKY AVENUE

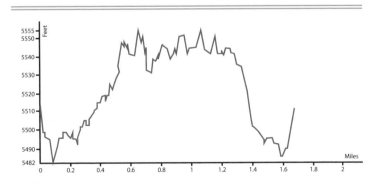

Feet

5555
5550
5540
5530
5520
5510
5500
5490
5482

Miles

0   0.2   0.4   0.6   0.8   1   1.2   1.4   1.6   1.8   2

DESCRIPTION OF THE ROUTE

Birdwatchers and history buffs will enjoy this easy hike around a neighborhood park that's just a stone's throw from Old 121 Brewhouse.

Nestled in the Denver suburb of Lakewood, this lovely urban park is just a couple blocks north of the Old 121 Brewhouse, so by tacking on a few steps on city sidewalks, you can literally hike to the doorstep of the brewery.

The site of Belmar Park once belonged to May Bonfils Stanton, heiress to the Denver Post fortune. During the Great Depression, Stanton had an exact replica of Marie Antoinette's Petite Trianon Palace built on the land. After Stanton's death in 1970, however, the mansion was stripped and demolished. Three years later, the city of Lakewood acquired the land and did its best to preserve it as a relatively undeveloped respite for local citizens from the growing encroachment of surrounding civilization.

Begin your hike at the "trailhead" at the intersection of W. Kentucky Avenue and Yarrow Street and head north. You'll soon cross the small creek bed of Weir Gulch before getting onto a horse trail that parallels a paved walkway heading west. This will eventually take you through tall grasses and open fields, up to the quaint Kountze Lake—a small remnant of what used to be a 50-acre manmade lake on Stanton's estate. Pause at the covered bridge on the east side of the lake to look for ducks, geese, hawks, egrets, belted kingfishers, swallows, and other birds nesting in the trees on the nearby island.

If you're facing the lake here, behind you lies the Lakewood Heritage Center, an optional detour if your curiosity is piqued. Thousands of historical artifacts are housed there, including a 1920s one-room schoolhouse, vintage farm machinery, rotating art exhibits, and restored remnants of the Bonfils estate including a calving barn and old auction booth. When you're ready to continue on around the lake, enjoy the cottonwood trees and views of neighboring foothills along your way. A lollipop loop takes you back to where you started.

If you happen to visit in late September or early October, you might be there in time for Cider Days, an annual festival in Belmar Park that celebrates Lakewood's agricultural history.

TURN BY TURN DIRECTIONS:

1. Head due north from the trailhead.
2. At the second junction, at 0.1 miles, make a left onto the dirt horse trail paralleling the sidewalk.
3. At 0.3, trail winds to the right. Cross the dirt road and continue north.
4. At 0.4, go right, then right again for a counterclockwise loop around the lake.
5. At 0.55, go left to the covered bridge and continue around lake.
6. At 0.8, go straight at trail junction.
7. At 1.1, go right to complete your lollipop loop.

FIND THE TRAILHEAD

This hike starts and finishes at the intersection of Yarrow Street and W. Kentucky Avenue, which is one block north and two blocks east of Old 121 Brewhouse in Lakewood.

OLD 121 BREWHOUSE

Old 121 Brewhouse opened its doors in March 2019 when a group of friends purchased their favorite local watering hole (formerly the home of Caution Brewing) and brought their many combined years of brewing and business experience to the neighborhood. In addition to their easy-drinking, classic-style beers, they also make their own craft sodas including a house root beer. Stay tuned to their events calendar to time your visit with a live music performance, bingo or trivia night, community talks or "Nama-Stay for a Drink"—a yoga class followed by a chance to enjoy a craft beer, soda or tea with your fellow yogis.

CONTACT INFORMATION
Lakewood Parks,
Forestry and Open Space,
480 S. Allison Parkway,
Lakewood, CO 80226;
303-987-7800

BREWERY/RESTAURANT
Old 121 Brewhouse
1057 S. Wadsworth Boulevard #60,
Lakewood, CO 80226
303-986-0311
Miles from trailhead: 0.3

DENVER LITTLETON

WANDER THE HIGH LINE CANAL TRAIL TO A BIRDWATCHER'S PARADISE

▷⋯ STARTING POINT	⋯✕ DESTINATION
WEST END OF W. BRANDON PLACE	**ROXBOROUGH COVE**
🍺 BEER	🁢 HIKE TYPE
HELLUVA CAUCASIAN STOUT	**URBAN WALK**
$ FEES	📅 SEASON
NONE BY THIS ROUTE	**YEAR-ROUND**
⛰ MAP REFERENCE	🐾 DOG FRIENDLY
CHATFIELD STATE PARK TRAIL MAP AT CPW.STATE.CO.US	**YES (LEASH REQUIRED)**
🕐 DURATION	↦ LENGTH
2 HOURS	**4 MILES**
↑↓ LOW POINT / HIGH POINT	〰 ELEVATION GAIN
5,450 FEET / 5,560 FEET	**MINIMAL**

 ALCOHOL 8.1% CONTENT   STOUT

 BLACK, MAHOGANY TINT

DARK CHOCOLATE, RAISINS, PEANUT BUTTER

ESPRESSO, SWEET ROASTY MALT

BITTERNESS
IBU: 30

SWEETNESS

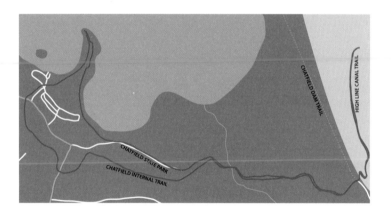

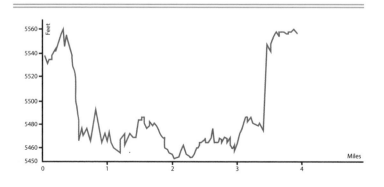

DESCRIPTION OF THE ROUTE

Live the dream with this relaxing jaunt featuring cottonwood trees, lake views, as well as beachcombing and bird-watching opportunities.

You wouldn't know it from the developed suburbs that surround them, but Chatfield State Park and the adjacent High Line Canal Trail (on which this lollipop loop begins and finishes) are steeped in history. The park is named for Isaac W. Chatfield, who originally traveled west to the Rockies by wagon train in search of gold before eventually becoming a decorated Civil War Lieutenant.

In 1975, the U.S. Army Corps of Engineers built the reservoir that lies at the heart of the state park. Originally engineered to help control flooding in the area, it now also serves as multipurpose water storage for the region. Skirting its shores to the east is the 71-mile High Line Canal Trail. The trail runs adjacent to its eponymous canal, an irrigation project developed after the gold rush in the 1850s to deliver water to new settlers and farmers in the area.

Today, a short stretch of the High Line Canal Trail serves as a perfect connector trail between the literal backyard of Living the Dream Brewing to Chatfield State Park. Just a hop, skip, and a jump over two sets of railroad tracks and you'll be immersed in dense cottonwood trees. The path is paved here as you meander through the woods before popping out in an open grasslands meadow. Listen for the hum of model airplanes in the park to your left; you might get treated to an aerial show.

As the path wraps back around toward the water, walk past the boat inspection hut and enjoy a stroll on the dirt path along the water. You'll pass a very small sandy beach and some picnic tables. Toward the end of the peninsula, keep an eye out for pelicans and other bird species that love to feed here.

In fact, the entire route is ideal for birdwatchers, as some 345 bird species have been recorded in the park. As you complete your lollipop loop and climb back up over the railroad tracks, hopefully your appetite will be thoroughly whetted for the rich, creamy White-Russian-inspired stout that has drawn the adoration of many a visitor to Living the Dream's outdoors-inspired taproom.

TURN BY TURN DIRECTIONS:

1. After a third of a mile on the High Line Canal Trail, cross both sets of railroad tracks, then go down steep gravel pitch to the Chatfield State Park sign at the entrance to the woods.
2. Follow paved path, winding left of parking area at 1 mile.
3. At miles 1.5 and 1.8, cross roads, then take path on right to go to water's edge.
4. Walk out to peninsula tip, then pick up path on other side to join dirt road.
5. Rejoin original trail at 2.9 and return the way you came.

FIND THE TRAILHEAD

The trailhead is just down the street from Living the Dream. Go north on N. Dumont Way, then make a left onto W. Brandon Place. Park here (or walk over from the brewery), then head south on foot past the sign for the High Line Canal Trail.

LIVING THE DREAM BREWING

Owners Carrie Knose and Jason Bell are avid backpackers and skiers whose love for the outdoors—and great beer—is abundantly evident in the mountain- and adventure-themed décor of their taproom. From the reclaimed wood paneling to the ski décor to the tree-stump rounds on which taster flights are served, Living the Dream Brewing is a haven for mountain lovers. Their bold, innovative brews (usually blessed with witty, pun-laden names) range from seasonals like a fruity Belgian ale ("War and Peace") or barrel-aged Festivus winter warmer to popular flagships like the Helluva Caucasian, a rich stout inspired by "The Dude" of *The Big Lebowski* and his love for White Russians.

CONTACT INFORMATION
Chatfield Sate Park,
11500 N. Roxborough Park Road,
Littleton, CO 80125;
303-791-7275

BREWERY/RESTAURANT
Living the Dream Brewing
12305 N. Dumont Way,
Littleton, CO 80125
303-284-9585
Miles from trailhead: 0.3

DENVER
MILE-HIGH LOOP

AN EASY, URBAN STROLL IN DENVER'S CITY PARK

▷⋯ STARTING POINT	⋯✗ DESTINATION
E 17TH AVENUE AND STEELE STREET	**DENVER CITY PARK**
🍺 BEER	HIKE TYPE
RARE TRAIT IPA	**URBAN WALK**
$ FEES	SEASON
NONE	**YEAR-ROUND**
⛰ MAP REFERENCE	🐾 DOG FRIENDLY
CITY PARK MAP AT DENVERGOV.ORG	**YES**
🕐 DURATION	↦ LENGTH
1-1.5 HOURS	**3.2 MILES**
↕ LOW POINT / HIGH POINT	〜 ELEVATION GAIN
5,272 FEET / 5,338 FEET	**MINIMAL**

 IPA

 CLOUDY DARK
YELLOW/ORANGE

PINEAPPLE, MANGO,
LEMON

ZESTY; TANGY
GRAPEFRUIT

BITTERNESS
IBU: 65

5
4
3
2
1

SWEETNESS

5
4
3
2
1

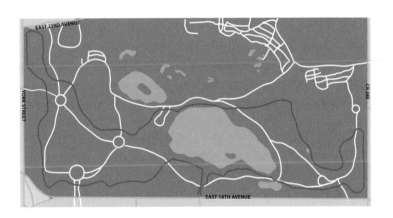

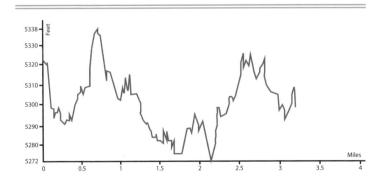

DESCRIPTION OF THE ROUTE

Celebrate the city's moniker, the "Mile-High City," with this urban loop a mile above sea level before grabbing a richly fruity IPA at nearby Cerebral Brewing.

City Park is the Central Park of Denver—an urban oasis of ponds, trees, fountains, trails, bicycle paths, and cultural sights, surrounded on all sides by a grid of residential neighborhoods and skyscrapers. One of the best views of the Denver skyline, silhouetted against the snow-capped Rocky Mountains to the west, is visible from the northeast side of the loop, which skirts the lawns of the city zoo and Denver Museum of Nature and Science.

Completed in 2009 and renovated more recently, the Mile-High Loop is a three-mile dirt path meandering loosely near the perimeter of the park. Its name is a nod to Denver's moniker of the "Mile-High City" for its 5,280-foot elevation above sea level. Trail markers exist in some places along the way, but not at all junctions—though it would be nearly impossible to get lost in the flat, 330-acre park. Just keep moving roughly around its perimeter, and you'll never be far from the official Mile-High Loop route.

As you ramble around the park, enjoy the shade of the massive oak trees. Ferril Lake, the central body of water in the park, is a popular nesting spot for herons, egrets, ducks, and Canada geese. Check the schedule at www.denverelectricfountain.org, but throughout summer evenings, you can expect to be treated to an electric light show on the lake. Paddleboard rentals are also available in the summer.

Two miles into your loop, at the northwest corner of the park, enjoy glimpsing the front of the Saint Ignatius Loyola Catholic Church across the street. Nearly a century old, the church's gorgeous, cathedral-esque twin towers are listed on the National Register of Historic Places. Several Civil War cannons are also located inside the park along this loop.

When you've exhausted your time at the park and its surrounding sights, head south on Monroe Street to reach the spacious taproom of Cerebral Brewing, which opened its doors in 2015. Hopheads will rejoice at its rotating selection of sessionable English- and Belgian-inspired styles, IPAs, saisons, and other unconventional takes on popular American styles.

TURN BY TURN DIRECTIONS:

1. Entering the park from Steele Street, cross the grass to get on the path and go right for a counter-clockwise loop.

2. At 0.7 miles, go left onto a dirt path, following signs for the Mile High Loop.

3. Stay left at an unmarked fork shortly after. At 1.15, keep following the dirt path now wrapping to your left.

4. At 1.3 and 1.5, cross paved roads, continuing on a dirt trail beyond the Mile-High Loop sign. Continue following Loop signs.

5. At 3.1, cross the grass to your right to return to your entry point, or continue four more blocks to Monroe Street, where Cerebral Brewing lies two blocks south.

FIND THE TRAILHEAD

The Denver City Park entrance at E. 17th Avenue and Steele Street within Denver city limits is accessible by walking, taking the Route 20 bus or driving and parking on nearby city streets. (Or the loop can be started at any other park trailhead with a parking lot.)

CEREBRAL BREWING

Located inside a former auto-body shop and headed up by biologists turned home brewers turned entrepreneurs, Cerebral Brewing is a beer aficionado's paradise masquerading as a hipstery haven for science geeks. Beer quality is monitored in an on-site lab; their logo is a human brain cleverly made of hops; taster glasses are chemistry beakers; the taproom's walls are decorated with science textbook pages. Just two blocks south of City Park, this is a perfect little neighborhood haunt for discerning hopheads and urban hikers alike.

CONTACT INFORMATION
Denver Parks and Recreation,
201 West Colfax Avenue, Dept. 601,
Denver, CO 80202;
720-913-1311

BREWERY/RESTAURANT
Cerebral Brewing
1477 Monroe Street,
Denver, CO 80206
303-927-7365
Miles from trailhead: 0.4

GOLDEN

TOUR THE TOP OF A 60-MILLION-YEAR-OLD MESA FORMED BY LAVA

▷⋯ STARTING POINT	⋯✗ DESTINATION
NEW TERRAIN BREWING COMPANY	**MESA TOP**
🍺 BEER	HIKE TYPE
CRUISE RIDE AMERICAN CREAM ALE	**MODERATE**
$ FEES	SEASON
NONE	**YEAR-ROUND**
⛰ MAP REFERENCE	🐾 DOG FRIENDLY
NORTH TABLE MOUNTAIN PARK MAP AT JEFFCO.US	**YES (LEASH REQUIRED)**
🕐 DURATION	↦ LENGTH
2-2.5 HOURS	**4.2 MILES**
↑↓ LOW POINT / HIGH POINT	〜 ELEVATION GAIN
5,656 FEET / 6,444 FEET	**780 FEET**

 CREAM ALE

 HAZY GOLDEN

BISCUITY; HONEY

TART LEMON; SMOOTH, CRISP

BITTERNESS
IBU: 13

SWEETNESS

5
4
3
2
1

5
4
3
2
1

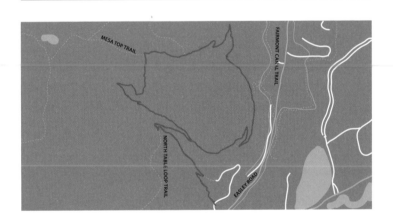

MESA TOP TRAIL

FAIRMONT CANAL TRAIL

NORTH TABLE LOOP TRAIL

EASLEY ROAD

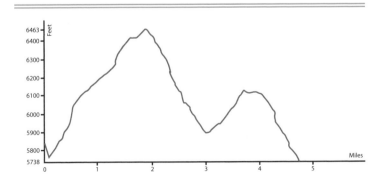

Feet

6463
6400
6300
6200
6100
6000
5900
5800
5738

Miles

0   1   2   3   4   5

DESCRIPTION OF THE ROUTE

Take a tour across the top of a 60-million-year-old mesa, formed by lava, that's now home to raptors, rattlesnakes, and panoramic views.

With bluffs and geologic formations beckoning to be explored, the 2,000-acre North Table Mountain Park looms over downtown Golden. Multiple trailheads lead to some 17 miles of trails set over the park's dry, desert-landscape mesa, so check out a map before you start. (And don't forget to lather on sunscreen!) This lollipop loop offers a good tour of the eastern side of the park, which tends to be a bit quieter than the western side.

The mesa that sits at the heart of the park was formed by lava flows thought to have occurred some 60 million years ago. To reach the expansive, flat top of the mesa, you'll climb up a dramatic drainage toward towering bluffs. The power lines over them are a bit of an eyesore amidst the otherwise pristine landscape, but the cliffs are still an impressive sight to behold. Soon you'll be atop them.

Keep an eye out overhead as you climb for raptors such as hawks, eagles, and falcons. At the same time, watch your step, for rattlesnakes are known to frequent the grasslands here. Indeed, the prairie serves as home to an impressive mix of flora and fauna in addition to the aforementioned, including deer, coyotes, cacti, cattail, and yarrow.

Many options exist for a longer excursion, including a loop on the Rim Rock Trail (closed February to July for raptor nesting) that tacks on about two miles atop the mesa. But if you choose to follow the turn-by-turn directions for this loop, you'll enjoy a brief tromp atop the mesa before dipping down singletrack on its east side. From here, enjoy the soaring views of the urban lakes and reservoirs below, and the Denver skyline farther off in the distance.

Watch for mountain bikers, as this park is popular for biking. In the summer, shadeless as they are, these trails can turn into a furnace in the middle of the day—so if you do go, get an early start to beat the heat. Otherwise, spring, fall, and even winter—depending on snow conditions—are all good seasons to take advantage of these trails.

TURN BY TURN DIRECTIONS:

1. From the west side of the brewery's parking lot, head north, then west, on the dirt path.
2. After a quarter-mile, go right to cross the bridge and road.
3. At 0.5 miles, go left on the trail. After several switchbacks, stay left to go clockwise on the North Table Loop.
4. At 1.5, stay right.
5. At 2.5, turn right onto the Mesa Top Trail.
6. At 3.1, turn right onto the North Table Loop.
7. At 3.5, reconnect with original junction. Go left to return the way you came.

FIND THE TRAILHEAD

This loop hike can be done from the parking lot of New Terrain Brewing Company, located on Table Mountain Parkway in Golden.

NEW TERRAIN BREWING COMPANY

With a motto like "For those who wander" and a location tucked away on the fringes of Golden's North Table Mountain Park, New Terrain Brewing Company is a no-brainer destination for all the hikers, trail runners, and mountain bikers who frequent the nearby trails. Opened by beer-industry professionals in 2016, the thirty-barrel brewhouse, taproom, and beer garden (complete with farm-to-fork food-truck cuisine) offer terrific views of the local foothills.

CONTACT INFORMATION
Jefferson County Open Space;
700 Jefferson County Parkway,
Golden, CO 80401;
303-271-5925

BREWERY/RESTAURANT
New Terrain Brewing Company
16401 Table Mountain Parkway,
Golden, CO 80403
720-697-7848
Miles from trailhead: 0

EVERGREEN

HIKE THE FOOTHILLS IN ALDERFER/THREE SISTERS PARK

▷⋯ STARTING POINT	⋯✕ DESTINATION
EAST TRAILHEAD, 30299 BUFFALO PARK ROAD	**EVERGREEN MOUNTAIN**
🍺 BEER	🎫 HIKE TYPE
IMPERIAL BLACK SAISON	**MODERATE**
$ FEES	📅 SEASON
NONE	**APRIL TO OCTOBER**
⛰ MAP REFERENCE	🐾 DOG FRIENDLY
ALDERFER THREE SISTERS PARK ON JEFFCO.US	**YES (LEASH REQUIRED)**
🕐 DURATION	↦ LENGTH
2.5-3.5 HOURS	**6 MILES**
↑↓ LOW POINT / HIGH POINT	〰 ELEVATION GAIN
7,513 FEET / 8,556 FEET	**990 FEET**

 IMPERIAL BLACK SAISON

ALCOHOL CONTENT 8.2%

 DARK BROWN

WOOD, YEAST

EARTHY, SPICY; BITTER FINISH

BITTERNESS
IBU: 30

SWEETNESS

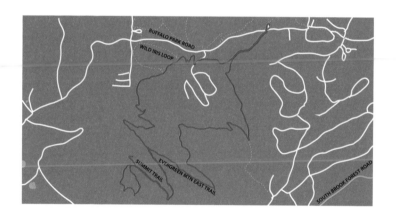

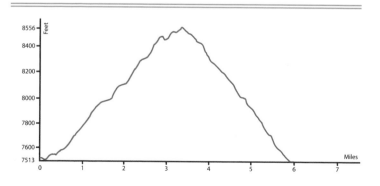

BUFFALO PARK ROAD
WILD IRIS LOOP
SUMMIT TRAIL
EVERGREEN MTN EAST TRAIL
SOUTH BROOK FOREST ROAD

DESCRIPTION OF THE ROUTE

Spend a day basking in Colorado sunshine on this terrific day hike in the Evergreen foothills, followed by an imperial saison on a scenic outdoor deck.

Alderfer/Three Sisters Park was once a working cattle ranch. These days, however, it's a public open space with some 15 miles of trails for hikers, mountain bikers, and equestrians to enjoy. The wildlife, flora, rock formations ideal for bouldering, historical structures, and dense, diverse forest offer a bit of everything for every kind of outdoors enthusiast.

The rocky outcroppings of "The Three Sisters" (North, Middle, and South) and "The Brother" dominate the skyline of the north side of the park, while Evergreen Mountain is the park's most prominent point on the south side. From its modest summit, it's possible to glimpse the Continental Divide and nearby 14,264-foot Mount Evans.

Many loop options exist to take you around either side of the park. This particular loop is a nice sampler platter of the park's terrain, leading you through nicely shaded lodgepole and ponderosa pine forest and cresting out at its high point atop Evergreen Mountain. However, you can also view the park trail map online at www.jeffco.us, or pick up a paper map at the trailhead and design your own loop.

In the late spring and early summer, wildflowers blossom abundantly. These are also popular seasons for trail users of all kind in this park, especially mountain bikers. Though the area receives ample snow in the winter, the trails are popular enough that the snow remains fairly packed— so it is possible to hike in the winter as well, and doing so will likely afford a more solitary hiking experience than in other seasons. (Wear traction devices on your shoes or boots if you're worried about slipping.)

After a few hours in the park, there's no need to head indoors; the expansive, sun-drenched back decks and dog-friendly beer garden at nearby Lariat Lodge Brewing await you.

TURN BY TURN DIRECTIONS:

1. From the East Trailhead, cross the road to pick up the Evergreen Mountain East Trail.
2. At the fork at 0.3 miles, stay left to initiate a clockwise loop.
3. At 2.2, go left and up the Summit Trail.
4. At 2.8, check out the scenic 0.1-mile spur before continuing on up to the 0.5-mile lollipop summit loop, which you can do in either direction.
5. On your way back down, make a left at the well-signed junction with Evergreen Mountain Trail West to complete a loop back to the trailhead.
6. Stay right at the next two intersections until you see a sign pointing you onto Ranch View Trail. Take that until you rejoin the Evergreen Mountain East Trail. Go left to return to the trailhead in 0.3 miles.

FIND THE TRAILHEAD

This is a very popular trailhead, so plan to arrive early. From downtown Evergreen, head west on Highway 73. Make a slight right onto Buffalo Park Road. In just over a mile, the East Trail parking lot for Alderfer/Three Sisters Park will be on your right, located at 30299 Buffalo Park Road.

LARIAT LODGE BREWING

The highlights of this brewery are its massive outdoor decks and beer garden—half of which is dog friendly—with expansive views of the surrounding mountains and woods. The inside is cozy and rustic as well, though. With the brewery's constantly rotating taps of classic beer styles and seasonal experiments, as well as a delicious menu (the Brussels sprouts salad is phenomenal), you'd be hard pressed to find a better place to refuel after your hike.

CONTACT INFORMATION
Jeffco Open Space,
700 Jefferson County Parkway,
303-271-5925

BREWERY/RESTAURANT
Lariat Lodge Brewing
27618 Fireweed Drive,
Evergreen, CO 80439
303-674-1842
Miles from trailhead: 2.9

MONUMENT

TROMP THROUGH WOODS AND GRASSLANDS
AT MONUMENT PRESERVE

▷··· STARTING POINT	···✗ DESTINATION
MT. HERMAN TRAILHEAD	**MONUMENT ROCK**
🍺 BEER	🀫 HIKE TYPE
DEVILS HEAD RED ALE	**EASY**
$ FEES	📅 SEASON
NONE	**YEAR-ROUND**
⛰ MAP REFERENCE	🐾 DOG FRIENDLY
WESTERN MAPS LLC, MONUMENT PRESERVE	**YES (LEASH REQUIRED)**
🕐 DURATION	↦ LENGTH
1-1.5 HOURS	**2.7 MILES**
↑↓ LOW POINT / HIGH POINT	〰 ELEVATION GAIN
7,011 FEET / 7,211 FEET	**200 FEET**

ALCOHOL
7.3%
CONTENT

RED ALE

ELEVATING THE CRAFT OF BEER

PIKES PEAK
Brewing Co

DEVILS HEAD
RED ALE

 HAZY REDDISH BROWN

FAINT BREAD,
LEMON RIND

ZESTY; STRONG PINEY
HOPS

BITTERNESS	SWEETNESS
IBU: 38	

5
4
3
2
1

5
4
3
2
1

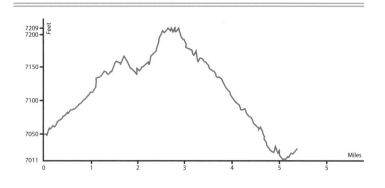

SCHILLING AVENUE

MOUNT HERMAN ROAD

7209
7200

7150

7100

7050

7011

Feet

Miles

0   1   2   3   4   5   5

DESCRIPTION OF THE ROUTE

The theme of this hike and beer pairing is fire! Tromp around on trails surrounding a USFS Fire Center before drinking a red ale dedicated to the area's wildlands firefighters.

This loop hike culminates in a visit to the striking rock formation from which the nearby town of Monument derived its name in the late 1800s. With its blend of shaded forest and open grasslands, it is especially ideal for spring and autumn mornings. (Due to its treeless exposure in places, it is not a good choice when lightning is present or imminent, so use caution when hiking in the summer.)

Monument Preserve—the area through which this hike lopes—is a unique place. It sits at the foot of Mount Herman, a 9,063-foot peak that dominates the western horizon from I-25, the main corridor between Denver and Colorado Springs. In addition to a number of singletrack trails open to hiking, mountain biking, and horseback riding, the area is also home to the Monument Fire Center. It is the summer home base for two different U.S. Forest Service wildlands fire operations teams, the Pike Hotshots and the Monument Helitack crew. For nearly six decades, the area also served as a National Forest System nursery, designed to help conduct reforesting efforts in areas decimated by wildfires or heavy logging. In the 1920s, a memorial grove was established to honor Rocky Mountain Region Forest Service employees who were killed during WWI, with individual trees planted for each service member lost.

Begin your hike with a relaxing ramble on quiet, wooded singletrack. You'll have already seen Monument Rock from the highway, but you'll get another glimpse of it as you pop out of the ponderosa and begin your approach through open fields of scrub oak. Enjoy a snack break at the small pond at the foot of Monument Rock. On the second half of your loop, continue straight through corridors of pine trees, following the trail left in front of the fence. Despite the complex web of trails in the Preserve, virtually no intersections are marked with any trail signs, so it is wise to carry a map.

When your hike is done, it's not far to get back to the town of Monument for a visit to Pikes Peak Brewing. This crisp, strong Devils Head red ale is dedicated to the very wildland fighters whose turf you've just spent the morning wandering around. It is named for the lookout tower atop Devils Head, the highest peak in the Rampart Range to the northwest of Monument, which provides a vantage spot for rangers to scope out wildfires, helping keep hikers and others out of danger's way.

TURN BY TURN DIRECTIONS:

1. Start on trail to the right of trailhead sign, and stay right at first few forks in the trail.
2. At mile 0.75, make a quick left on dirt road, then immediate right on trail just after the gate.
3. At 1.25, go left in front of the pond and arrive at Monument Rock.
4. Three trails branch off from east side of the rock. Facing away from the rock, take the one on your right.
5. At mile 2.5, go left at junction, then left again, then right under powerlines. Pass through a gate in an old barbed wire fence and go right to return to the trailhead.

FIND THE TRAILHEAD

Drive west out of Monument on 2nd Street, then make a left onto Mitchell Road. In a half-mile, turn right onto Mt. Herman Road. In 0.8 miles, make a left onto Nursery Road, and the trailhead will be on your right.

PIKES PEAK BREWING

Pikes Peak Brewing is a community-focused brewpub whose stated goal is to "elevate the craft of beer" while staying faithful to classic styles. Their year-round lineup (all available in cans, too) includes an IPA, red ale, pale ale, English mild, Belgian ale, and an oatmeal stout. Seasonal beers are often given hiking-related names, too, including a spiced holiday ale named for a local club that hikes Pikes Peak on New Years Eve each year to shoot off fireworks, and a one-time-only "Incline Imperial IPA," boasting a hearty 185 IBU and named for the nearby Incline in Manitou Springs.

CONTACT INFORMATION
U.S. Forest Service,
Pikes Peak Ranger District,
601 S Weber Street,
Colorado Springs, CO 80903;
719-636-1602

BREWERY/RESTAURANT
Pikes Peak Brewing
1756 Lake Woodmoor Drive,
Monument, CO 80132
719-208-4098
Miles from trailhead: 2.4

COLORADO SPRINGS

VISIT A 250-FOOT WATERFALL IN NORTH CHEYENNE CAÑON PARK

▷··· STARTING POINT	···✕ DESTINATION
GOLD CAMP ROAD	**ST. MARY'S FALLS**
🍺 BEER	🔣 HIKE TYPE
CHEYENNE CAÑON PIÑON NUT BROWN ALE	**MODERATE-STRENUOUS**
$ FEES	SEASON
NONE	**YEAR-ROUND**
⚠ MAP REFERENCE	🐾 DOG FRIENDLY
POCKET PALS TRAIL MAP #3: NORTH CHEYENNE CAÑON AREA	**YES (LEASH REQUIRED)**
⏱ DURATION	⊢ LENGTH
3-4 HOURS	**6 MILES**
↑↓ LOW POINT / HIGH POINT	〰 ELEVATION GAIN
7,533 FEET / 8,950 FEET	**1,400 FEET**

BROWN ALE

 CLEAR AMBER

 BREAD, COCOA

 NUTTY; CHERRIES

BITTERNESS | **SWEETNESS**
IBU: 25

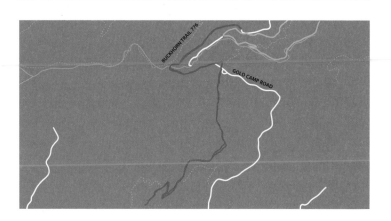

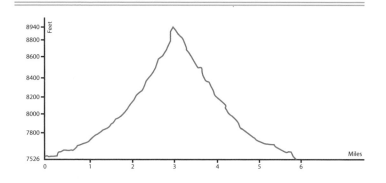

DESCRIPTION OF THE ROUTE

There's no better way to celebrate completing this popular waterfall hike than with a roasted piñon nut brown ale that celebrates and supports North Cheyenne Cañon Park.

Once a popular site for Ute Indian tribes, North Cheyenne Cañon Park today is an enchanted labyrinth of trails, waterfalls, rock formations, historic stone bridges, and other remnants of Colorado's gold rush days.

The first part of the hike follows the decommissioned Upper Gold Camp Road, first traversing the north side of a gully before making a 180-degree turn to traverse the south side. The gravel path here is wide, relatively flat, non-technical, and exposed to the sun. In just over a mile, you'll come to the gated entrance of an old tunnel. It was originally constructed in the 1800s during the gold rush as part of a railway before being converted to a toll road for cars. In the late 1980s, the tunnel collapsed and was never repaired, thereby shutting down vehicular access to Upper Gold Camp Road—and creating infinite fodder for Springs locals' ghost stories and haunted tales of paranormal occurrences in the creepy, fenced-off tunnel.

After you bypass the tunnel, the trail narrows to singletrack into an area called Buffalo Canyon. Shortly, look to your right for a sign marking the way to St. Mary's Falls. From here, the route is a fairly gentle climb up through mixed pine and spruce forest and along the babbling Buffalo Canyon Creek. A few minor trail junctions exist along the way, though most routes quickly reconnect. (Just ignore the offshoot trails on your left that go all the way down to the creek.)

Only the final third of a mile offers significant steepness as you approach the falls on a series of switchbacks. Enjoy watching the lovely, 250-foot cascade tumbling down the imposing granite wall above you. Depending on what time of year you go, the falls may be roaring (spring), more of a trickle (fall), or frozen altogether (winter).

Though this hike is doable year-round, if you go between November and April, snowshoes or traction devices and trekking poles are advised, as areas can be snow-covered or slick with ice.

On your drive back toward the Springs for your post-hike beverage, pull off about a half-mile down N. Cheyenne Canyon Road to check out the magnificent Helen Hunt Falls just off the road.

TURN BY TURN DIRECTIONS:

1. Start hiking on the dirt road at the back (northwest corner) of the parking lot.
2. After passing the tunnel, go right at mile 1.2, following the sign for St. Mary's Falls.
3. At 2.6, make a sharp right in front of the St. Mary's Falls sign and follow the trail up.
4. At 2.8, stay left and reach the falls in another quarter of a mile.

FIND THE TRAILHEAD

Drive southwest out of Colorado Springs on Cheyenne Boulevard, then make a slight right onto N. Cheyenne Canyon Road. Drive 3 miles to the end of the road and park in the gravel parking lot at the intersection with High Drive and Gold Camp Road.

BRISTOL BREWING COMPANY

Bristol Brewing Company is located inside the old Ivywild School, an elementary school built in 1916 and converted in 2009 into a multi-use community space filled with art, food, music, and, of course, a brewery. This approachable summer seasonal is made with real roasted pinyon pine nuts. Not only does it taste delicious, 100% of its profits directly support the trails in North Cheyenne Cañon Park by going to the nonprofit organization Friends of Cheyenne Cañon.

CONTACT INFORMATION
U.S. Forest Service,
Pikes Peak Ranger District,
601 S. Weber Street,
Colorado Springs, CO 80903;
719-636-1602

BREWERY/RESTAURANT
Bristol Brewing Company
1604 S. Cascade Avenue,
Colorado Springs, CO 80905
719-633-2555
Miles from trailhead: 5.8

MANITOU SPRINGS

HIKE THROUGH DRAMATIC SANDSTONE FORMATIONS

▷··· STARTING POINT	···✕ DESTINATION
RIDGE ROAD TRAILHEAD	**RED ROCK CANYON**
🍺 BEER	🔳 HIKE TYPE
HIGH GROUND IPA	**MODERATE**
$ FEES	📅 SEASON
NONE	**YEAR-ROUND**
⛰ MAP REFERENCE	🐾 DOG FRIENDLY
RED ROCK CANYON MAP AT COLORADOSPRINGS.GOV	**YES (LEASH REQUIRED)**
🕐 DURATION	↦ LENGTH
2 HOURS	**3.4 MILES**
↑↓ LOW POINT / HIGH POINT	〰 ELEVATION GAIN
6,191 FEET / 6,627 FEET	**430 FEET**

ALCOHOL 6.9% CONTENT

IPA

CLEAR, BRIGHT GOLDEN

GRASS, GRAPEFRUIT

PINE; ZESTY, HOP-CENTRIC

BITTERNESS
IBU: 125

SWEETNESS

Manitou Brewing Co.

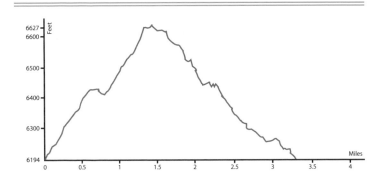

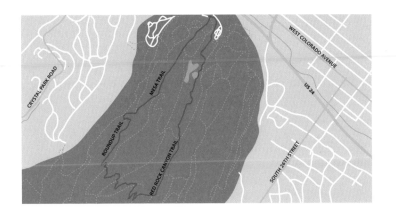

DESCRIPTION OF THE ROUTE

Red Rock Canyon is a less touristy alternative to the nearby, infamous Garden of the Gods, yet features many of the same dramatic sandstone formations.

This loop hike offers a solid tour of some of Red Rock Canyon's spunkiest singletrack, scenic ridges, eye-popping geologic formations, and a lovely, quiet pond—watch for geese and herons—with which to wrap up your hike. A number of trails exist in this park, so the loop suggested here is merely that—a suggestion. Plenty of other routes exist, too; your best bet is to carry a map and explore to your heart's delight.

True to the park's name, the crushed-gravel trails wind through many a red-rock landscape. The rock here is primarily sandstone of the Fountain Foundation—formed some 300 million years ago by the erosion of the Ancestral Rocky Mountains. The same Fountain Foundation composes Boulder's iconic Flatirons, nearby Garden of the Gods, and Denver's famous Red Rocks Amphitheater. It gets its color from the copious pink feldspar grains embedded in it.

This hike takes you through a mix of Douglas fir and ponderosa pine forest in the shadier stretches of the canyon, as well as more desert-like environs replete with Gambel oak, pinyon and juniper trees, yucca, and prickly pear. Keep an eye out for the rustle of birds, deer, coyotes, or other smaller mammals in the bushes around you as you hike, as the area is filled with vibrant wildlife. And, of course, soak up all the impressive views along the way of the park's namesake red-rock formations.

If you bring Fido with you, you can also make use of the off-leash dog area (include a short loop trail in addition to those included in this hike) just west of the Mesa Trail, shortly after you begin hiking. Either way, when you're ready to venture out onto the loop, stay on the Mesa Trail and enjoy the views of Pikes Peak on the horizon to your right before getting on the Roundup Trail and returning on the Red Rock Canyon Trail. At times along this hike, you'll also be able to catch glimpses of the slanted red rocks of Garden of the Gods to the north.

Though nearly 100 technical climbing routes have been established in the park for permitted climbers to tackle, be aware that rock scrambling more than 10 feet off the ground (sans permits, ropes, and harnesses) is illegal in this area, and punishable by fines and/or jail time.

TURN BY TURN DIRECTIONS:

1. Start up Mesa Trail from end of the parking area.
2. At mile 0.55, make a right on Quarry Pass, following it until mile 0.8, when you'll wrap around red rocks onto the Roundup Trail.
3. At 1.2, follow a switchback around to the left, staying on Roundup Trail.
4. At mile 2, at the bottom of the hill, go left, then left again in a tenth of a mile onto Red Rock Canyon Trail.
5. At mile 3, make a slight left (and stay left) to return to your car.

FIND THE TRAILHEAD

Red Rock Canyon Open Space is located along Highway 24 at 3550 W. High Street. From Manitou Springs, drive east on Highway 24, then make a right on to Ridge Road and immediate left (via roundabout) onto High Street. Park in the first lot on your right.

MANITOU BREWING COMPANY

Manitou Brewing Company is located in a former burro stable, where hikers in the late 1800s who wanted to summit Pikes Peak could rent a donkey to aid with their journey. Today, it is a thriving spot for both craft beer and craft food, with beautiful repurposed wooden décor, a scrumptious menu, and a wide variety of house beers and guest taps alike. This High Ground IPA is dry hopped with more than five pounds of hops per barrel, and earned a World Beer Cup Silver Medal in 2016.

CONTACT INFORMATION
Colorado Springs Parks,
Trails & Open Spaces,
1401 Recreation Way,
Colorado Springs, CO 80905;
719-385-5940

BREWERY/RESTAURANT
Manitou Brewing Company
725 Manitou Avenue,
Manitou Springs, CO 80829
719-282-7709
Miles from trailhead: 2

PUEBLO

WANDER A DESERT-PRAIRIE MESA AT LAKE PUEBLO STATE PARK

▷··· STARTING POINT	···✕ DESTINATION
ARKANSAS POINT CAMGPROUND	**KEYHOLE CANYON**
🍺 BEER	🔲 HIKE TYPE
IRISH RED ALE	**EASY-MODERATE**
$ FEES	📅 SEASON
$8	**YEAR-ROUND**
🗺 MAP REFERENCE	🐾 DOG FRIENDLY
LAKEPUEBLOTRAILS.ORG	**YES (LEASH REQUIRED)**
🕐 DURATION	↦ LENGTH
2 HOURS	**4.7 MILES**
↑↓ LOW POINT / HIGH POINT	〰 ELEVATION GAIN
4,902 FEET / 5,164 FEET	**255 FEET**

5.0% ALCOHOL CONTENT

RED ALE

DEEP AMBER

TOASTY MALTS, BREADY NOTES

CRISP; FAINT CARAMEL

BITTERNESS
IBU: UNLISTED

SWEETNESS

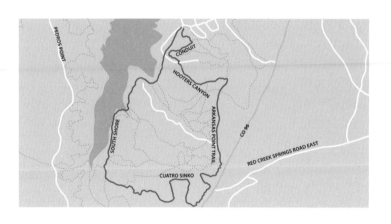

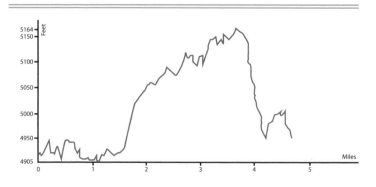

DESCRIPTION OF THE ROUTE

Take advantage of one of Colorado's sunniest cities by soaking up rays and shimmering lake views on a desert-prairie mesa at Lake Pueblo State Park.

For gorgeous desert plains wandering, it's hard to beat the serene environs of Lake Pueblo State Park. Thanks to its off-the-beaten-path location and the park's reputation as more of a boater's paradise than a hiker's destination—as of the writing of this book, the park's trail map doesn't even include the majority of the trails that compose this loop—you're likely to have the singletrack largely to yourself.

This loop utilizes the South Shore Trail, Rock Canyon, Keyhole Canyon, Cuatro Sinko, Arkansas Point, Hooters Canyon, and Conduit trails for a peaceful hike that provides a bit of everything—lovely views of the reservoir framed by larger mountain ranges in the distance, rock formations, twisting canyons, abundant botanical life on the short-grass-prairie mesa (including numerous wildflowers, cacti, rice grass, yucca plants, winterfat, and more) and possible wildlife sightings ranging from mule deer and bobcat to great blue herons and bald eagles. Keep an eye out above for majestic raptors circling in the sky, as well as an eye on the ground around you for snakes, including the occasional rattler.

This route stays relatively mellow, primarily remaining on smooth and flat or gently rolling singletrack the whole way. Though it would be difficult to get too turned around here given the openness of the area, it is not a bad idea to have a map on you—especially if you're interested in checking out any of the additional trails here on the south side of the park (of which there are many), particularly the various gullies and rock canyons that cut horizontally across this trail system.

Toward the end of this loop, you'll climb up to a nice bench lookout with 180-degree views of the lake, nearby rock formations, and surrounding mountain ranges. It's a perfect place to catch a sunrise or sunset before returning to downtown Pueblo for some hearty Irish grub and craft beer. Or, pitch a tent at Arkansas Point campground, literally a few steps away from the trailhead, and save your brew sampling for the next day.

TURN BY TURN DIRECTIONS:

1. Facing south, start on clear singletrack between the trail to your left and dirt road to your right.
2. Follow the South Shore Trail, staying right (hugging the water) at numerous trail junctions.
3. At 1.3 miles, stay left (on South Shore Trail) at junction with Creekside.
4. At 1.4, veer left onto Rock Canyon, then right at 1.6 onto Keyhole Canyon.
5. At 2.1, go left (at unmarked junction) onto Cuatro Sinko.
6. At 3.1, make a slight (not sharp) left onto Arkansas Point Trail.
7. At 3.75, turn left onto Hooters.
8. Cross dirt road at 4.2 to continue onto Conduit Trail. Make a left at 4.5 to return to trailhead.

FIND THE TRAILHEAD

From downtown Pueblo, drive west out of town on W. 4th Street/Thatcher Road/Highway 96. Turn right onto S. Marina Road into Lake Pueblo State Park, then follow signs for Arkansas Point Campground. Park by the covered kiosk, before the campground.

SHAMROCK BREWING COMPANY

Part craft brewery, part Irish pub, Shamrock offers a friendly atmosphere for enjoying delectable Irish-inspired creations. Brewed with local Colorado malts, their beers range from sessionable house brews to seasonal experiments like a 9.2% ABV "Big PAPA" imperial IPA and a coffee stout collaboration with neighborhood Pueblo coffee roasters Solar Roast. And, if you've never tried Irish boxty, now's the time.

CONTACT INFORMATION
Colorado Parks and Wildlife,
Pueblo Field Office,
600 Reservoir Road,
Pueblo, CO 81005,
719-561-5300

BREWERY/RESTAURANT
Shamrock Brewing Company
108 W. 3rd Street,
Pueblo, CO 81003
719-542-9974
Miles from trailhead: 8.2

TRINIDAD

DOUBLETRACK DELIGHT AT TRINIDAD LAKE STATE PARK

▷⋯ STARTING POINT	⋯✕ DESTINATION
SOUTH SHORE CAMPGROUND	**LONG'S CANYON**
🍺 BEER	🗺 HIKE TYPE
MILK STOUT	**EASY-MODERATE**
$ FEES	📅 SEASON
$8	**YEAR-ROUND**
🗺 MAP REFERENCE	🐾 DOG FRIENDLY
TRINIDAD LAKE STATE PARK AT CPW.STATE.CO.US	**YES (LEASH REQUIRED)**
🕐 DURATION	↦ LENGTH
2-2.5 HOURS	**5 MILES**
↑↓ LOW POINT / HIGH POINT	〰 ELEVATION GAIN
6,253 FEET / 6,319 FEET	**120 FEET**

 MILK STOUT

 BLACK; FOAMY HEAD

BANANA, CLOVES

 ROASTY, DARK FRUITS;
FULL-BODIED

BITTERNESS   **SWEETNESS**
IBU: 34

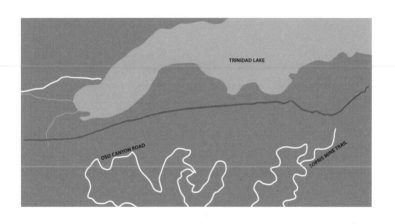

TRINIDAD LAKE

OSO CANYON ROAD

SOPRIS MINE TRAIL

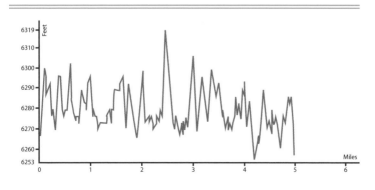

DESCRIPTION OF THE ROUTE

Nestled in the foothills of Colorado's stunning Sangre de Cristo Mountains, Trinidad has everything from mining history to scenic trails to an excellent local microbrewery.

Trinidad sits at the southeastern terminus of the Highways of Legends (Highway 12), a 70-mile Colorado Scenic Byway winding through the San Isabel National Forest with dozens of scenic vistas and historical or cultural points of interest along the way. Just 13 miles north of the New Mexico border, Trinidad also sits along the historic Santa Fe Trail. As such, it is an old mining town steeped in history. In fact, until the 1950s, it was one of Colorado's most populous towns, thanks almost entirely to its position along layers of Cretaceous period coal seams. Most of the town's coal mines closed, however, after World War II. Today, downtown Trinidad maintains a memorial to honor those who labored in the coal camps, while nearby Cokedale (seven miles west) stands as one of the state's few remaining examples of an intact coal camp.

Trinidad is also home to Trinidad Lake State Park, where you can generally count on enjoying some solitude on the park's doubletrack South Shore Trail. As its name suggests, the park also boasts an 800-acre lake, which is stocked year-round for fishing everything from rainbow trout and crappie to largemouth bass and channel catfish.

Your hike begins near the South Shore Campground and runs in a fairly straight line along the bluffs south of Trinidad Lake. From here, enjoy a great view of the reservoir, set against the dramatic backdrop of the Sangre de Cristo Mountains. The trail has a few gently rolling hills and offers a relaxed hike. You'll cut a wide path through the surrounding scrub oak, grama, and rabbitbrush, as well as pinyon and juniper trees; if you hike this around October and get lucky, you might even be able to harvest some pine nuts from the cones of the surrounding pinyon trees along the way. In the meantime, watch out for snakes basking in the sun on the trail. Keep an eye out, too, for deer, coyote, fox, raptors, great blue heron, and even several species of bats, which make their home amidst the woodlands and abandoned mine shafts within the park's boundaries.

At the end of the 2.5-mile South Shore Trail, you have the option to turn around come back the way you came, or add on another 1.5 miles (making for a 6.5-mile roundtrip hike) by veering left into Long's Canyon. Here, a 0.75-mile self-guided nature trail leads to several observation decks overlooking wetlands where you can continue your wildlife-watching adventures.

TURN BY TURN DIRECTIONS:

1. Start up the gravel path on the south side of the road, just past the campground.

2. Reach the end of the South Shore Trail at 2.5 miles. Turn around here for an easy out-and-back, or tack on some bonus hiking by veering left to explore Long's Canyon.

FIND THE TRAILHEAD

Take I-25 south from Trinidad. Take exit 11 and continue south on County Road 69.1. Make a slight right onto County Road 69, then another immediate right onto County Road 18.3. In 1 mile, turn left to enter State Park, then stay left at first fork, following signs for South Shore Campground. Drive past the campground and find the South Shore Trail shortly after on your left-hand side.

DODGETON CREEK BREWING COMPANY

This slightly out-of-the-way gem is well worth the trip (though check their hours before you go). An independent craft microbrewery and friendly neighborhood watering hole, Dodgeton Creek has a cozy tasting room featuring 10 taps with mostly mainstay brews and a few rotating, delectable seasonals at any given time—think Mexican Chocolate Porter, Pumpkin Stout, or India Pale Lager. A conservation-focused brewery, Dodegeton Creek also uses recycling, reclaimed resources, and composting to help minimize their environmental impact.

CONTACT INFORMATION
Colorado Parks and Wildlife,
Trinidad Lake State Park,
32610 Highway 12,
Trinidad, CO 81082;
719-846-6951

BREWERY/RESTAURANT
Dodgeton Creek Brewing Company
36730 Democracy Drive,
Trinidad, CO 81082
719-846-2339
Miles from trailhead: 10.5

DEL NORTE

A QUICK HIKE TO 360-DEGREE VIEWS OF THE SAN LUIS VALLEY

▷··· STARTING POINT	···✕ DESTINATION
COLUMBIA TRAILHEAD	**LOOKOUT MOUNTAIN**
🍺 BEER	🔲 HIKE TYPE
HOP TRASH IPA	**MODERATE**
$ FEES	📅 SEASON
NONE	**APRIL TO NOVEMBER**
⛰ MAP REFERENCE	🐾 DOG FRIENDLY
DELNORTETRAILS.ORG	**YES**
🕐 DURATION	↦ LENGTH
45 MINUTES	**1 MILE**
↕ LOW POINT / HIGH POINT	〰 ELEVATION GAIN
8,000 FEET / 8,475 FEET	**475 FEET**

IPA

COPPER

FLORAL; CITRUS

PINE HOPS; BITTER FINISH

BITTERNESS
IBU: 85

SWEETNESS

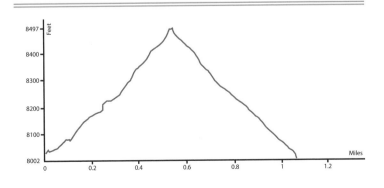

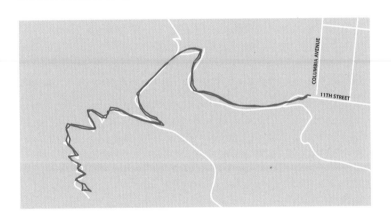

DESCRIPTION OF THE ROUTE

Just down the street from the brewery, this quick peak-bagging jaunt takes you up through desert grasses and wildflowers for a 360-degree view of the San Luis Valley.

In the late 1800s, the 8,475-foot summit of Del Norte's Lookout Mountain served as home to a massive telescope and domed observatory—at its time, one of the most powerful telescopes in the American West. Given Del Norte's high elevation and frequently clear night skies, it was perfectly situated for stargazing and astronomy research. Unfortunately, the telescope fell into disrepair by the end of the century, and by the mid 1900s, the observatory had been destroyed by storms and vandalism.

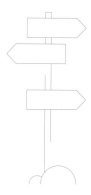

Today, the summit of Lookout Mountain makes for a lovely, quick hike straight from town. It is known for its large, white "D" monogram on its hillside, which can be seen for miles in many directions. You'll curl around it on your way up the mountain. The trail itself is a wellspring of yucca, rabbitbrush, sagebrush, prickly pear cactus, Indian paintbrush, and other wildflowers. There's not a lick of shade along the way, so wear a hat and sunscreen.

After ascending the final switchbacks to the summit, have a seat on the rocks (after checking for rattlesnakes) and admire the sprawling San Luis Valley all around you. You'll have a great view of town and the nearby Sangre de Cristo Mountains. Soak it all in before turning back down to make your way to Three Barrel Brewing, where a nice shaded outdoor terrace and an assortment of crisp, refreshing beers awaits—including their classic Hop Trash IPA, which features six different high-acid hops.

There is a trail map at the Columbia trailhead you can photograph for reference, but you'd have to work extraordinarily hard to get lost on this tree-less foothill with a view of town as a constant reference. There are further trails within Lookout Mountain Park if you'd like to extend your hike. Keep an eye out for rattlesnakes, and do not attempt this hike if lightning is present or forecast.

TURN BY TURN DIRECTIONS:

1. At 0.2 miles, go left at the signed junction, following sign to Summit Trail.
2. At 0.25, stay straight on main trail.
3. At 0.3, make a right at wooden post to start switchbacks up.
4. At 0.33, go left at the intersection.
5. At 0.5 miles, reach summit.

FIND THE TRAILHEAD

On the south end of town, the Columbia Trailhead can be found at the intersection of S. Columbia Avenue and 11th Street (two blocks east and six blocks south of the brewery).

THREE BARREL BREWING

Three Barrel Brewing is tucked away in the picturesque San Luis Valley between two majestic mountain ranges—the Sangre de Cristos and the San Juans. Its small-batch, artisanal ales are made with mountain water, fresh Colorado-grown malts, hops, honey, and other locally sourced ingredients. Drop by for wood-fired pizza and a broad assortment of lagers and ales—and some of the best brewery and beer-bottle artwork to be found in the state of Colorado.

CONTACT INFORMATION
Del Norte Trails Organization,
PO Box 309,
Del Norte, CO 81132;
719-657-2827

BREWERY/RESTAURANT
Three Barrel Brewing
475 Grand Avenue,
Del Norte, CO 81132
719-657-0681
Miles from trailhead: 0.5

ALAMOSA

GREAT SAND DUNES NATIONAL PARK'S OTHERWORLDY LANDSCAPE

▷⋯ STARTING POINT	⋯✗ DESTINATION
DUNES PARKING AREA	**HIGH DUNE**
🍺 BEER	🔠 HIKE TYPE
VALLE ESPECIAL MEXICAN LAGER	**VERY STRENUOUS**
$ FEES	📅 SEASON
$15 (OR ANNUAL PASS)	**YEAR-ROUND**
⛰ MAP REFERENCE	🐾 DOG FRIENDLY
GREAT SAND DUNES NATIONAL PARK VISITOR GUIDE AT NPS.GOV	**YES (LEASH REQUIRED)**
⏲ DURATION	⊢ LENGTH
2 HOURS	**2.6 MILES**
↑↓ LOW POINT / HIGH POINT	∿ ELEVATION GAIN
8,071 FEET / 8,720 FEET	**670 FEET**

  MEXICAN LAGER

 GOLDEN

FAINT CITRUS, APPLES

LIGHT-BODIED, GRAINY

BITTERNESS
IBU: UNLISTED

SWEETNESS

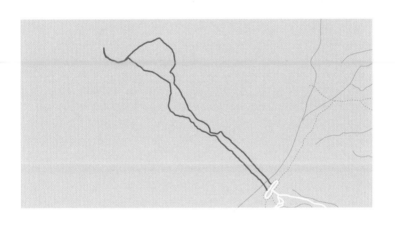

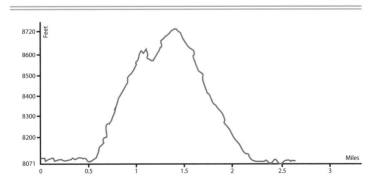

DESCRIPTION OF THE ROUTE

Get a dose of desert dunes in this otherworldy landscape—trekking up North America's tallest sand dunes, nestled at the foot of the jagged Sangre de Cristo Mountains.

Great Sand Dunes National Park is home to the tallest sand dunes on the continent. The dunes in this one-of-a-kind park are estimated to be nearly half a million years old and have remained surprisingly unchanged for at least the last century—which is rather amazing, given that they're essentially just massive mounds of sand, originally deposited by volcanic activity and resculpted daily by the whims of the wind.

For reasons that become obvious when you see how quickly wind erases even the deepest footsteps in the sand, there are no formal trails through the dune field. However, many park visitors delight in climbing High Dune, the most prominent mound on the eastern ridge of the dune field. It rises nearly 700 feet above the San Luis Valley. Climbing to its summit on soft sand (more than 8,000 feet above sea level to boot) is no easy feat—but the payoff is huge.

It's a "choose your own adventure" route to the top of High Dune, with many options for wending your way among the sculpted dunes and dramatic, curved ridgelines along the way. For an extra dose of fun and adventure, rent a sand board or sand sled at Great Sand Dunes Oasis (near the park entrance) to allow for a fast, adrenaline-filled ride back down.

Do beware that sand surface temperatures by midday in the summer can reach a scorching 150 degrees Fahrenheit. Do not venture out into the dunes without proper shoes (not sandals) to protect your feet—and, if you're hiking with a dog, not without booties to protect your pup's paws as well. Your shoes will inevitably fill with sand as you hike; if the sand temperature permits, it can be more enjoyable to hike barefoot, but always carry proper footwear with you in case sand temperatures heat up while you're out. Also beware of developing thunderstorms, as it's very dangerous to be in the dunes when lightning is present.

Of all the hikes in this book, this one is located the farthest away from its accompanying brewery—but the drive is worth it to visit this wildly unique landscape. You will feel as though you've arrived on a different planet from the rest of Colorado, let alone the rest of the continent.

TURN BY TURN DIRECTIONS:

1. Set your sights on the highest dune on the ridgeline visible from the dunes parking area.
2. Choose your own route to the top and back.

FIND THE TRAILHEAD

From Alamosa, take US-160 east out of town for 14 miles. Turn left (north) onto CO-150 and drive another 19 miles. The Dunes Parking Area is just past the visitor center.

SAN LUIS VALLEY BREWING COMPANY

After a hot day in the dunes, the light-bodied brews at family-owned San Luis Brewing Company hit the spot. The Valle Especial Mexican-style lager is served with a slice of lime—or, if you're in the mood for a little more heat, order a Valle Caliente for a green-chili infused rendition. The brewpub, which opened in 2006, is located in a gorgeously restored building downtown on Main Street. Head brewers Scott and Angie Graber also roast their own coffee beans, available for purchase on site.

CONTACT INFORMATION
Great Sand Dunes National Park
and Preserve,
11999 Highway 150,
Mosca, CO 81146;
719-378-6300

BREWERY/RESTAURANT
San Luis Valley Brewing Company
631 Main Street
Alamosa, CO, 81101
719-587-2337
Miles from trailhead: 34

CRESTONE

MAJESTY AND MYSTICISM IN THE "SHANGRI LA" OF THE ROCKIES

▷⋯ STARTING POINT	⋯✕ DESTINATION
WILLOW LAKE TRAILHEAD	**WILLOW LAKE**
🍺 BEER	🁢 HIKE TYPE
EMERGENCE AMERICAN PALE ALE	**MODERATE-STRENUOUS**
$ FEES	📅 SEASON
NONE	**JULY TO SEPTEMBER**
⛰ MAP REFERENCE	🐾 DOG FRIENDLY
TRAILS ILLUSTRATED 138: SANGRE DE CRISTO MOUNTAINS	**YES (LEASH REQUIRED)**
🕐 DURATION	↦ LENGTH
5-7 HOURS	**9 MILES**
↑↓ LOW POINT / HIGH POINT	〰 ELEVATION GAIN
8,921 FEET / 11,634 FEET	**2,735 FEET**

 PALE ALE

 CLOUDY AMBER

 FRUITY, ZESTY HOPS

 LIGHT, CRISP, SUBTLE MALT

BITTERNESS
IBU: UNLISTED

SWEETNESS

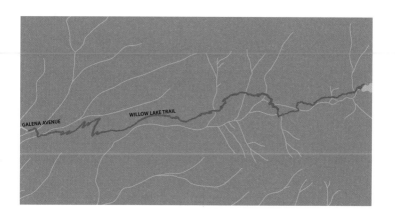

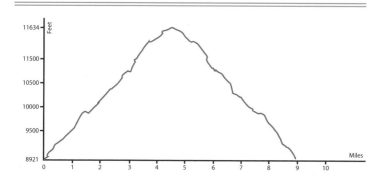

DESCRIPTION OF THE ROUTE

Soak up the majesty and mysticism of what's been called the "Shangri La" of the Rockies with this gorgeous lake hike and some of the state's best craft beer and food.

If you've never been to Crestone before, buckle up for a wild, mystical journey. Aside from being located in the heart of the mountains, this tiny town and the surrounding flatlands are home to dozens of spiritual organizations, ashrams, monasteries, stone labyrinths, temples, churches, healing centers, meditation spaces, and other sacred sites.

Though known primarily for its spiritual offerings, Crestone is also a below-the-radar hikers' and climbers' paradise. Nestled at the foot of Colorado's largest mountain range, the Sangre de Cristos (Spanish for "blood of Christ"), Crestone sits in the shadow of a handful of looming 14,000-foot peaks, including Crestone Peak, Crestone Needle, Kit Carson, Challenger Point, and Humboldt Peak. To their west lies the expansive, pancake-flat San Luis Valley.

The Willow Lake trail, which starts just a couple miles up a dirt road east out of town, is a wonderful introduction to some of the stunning, rugged backcountry that laps at the edges of Crestone. The trail, though often technical in footing, climbs at a manageable, steady grade; it presents a moderate challenge, but is rarely soul crushing in its steepness. Along the way, waterfalls, wildflowers, and creek crossings abound.

The trail begins with some soft sand in places before quickly giving way to firmer footing on hard-packed dirt and switchbacking up through shady fir and pine forest. Climb steadily for a little over a mile before the trail crests with open views of the large peaks ahead, then mellows with a brief downhill interlude to let you catch your breath. The steepest switchbacks arrive around 3.3 miles in, zig-zagging up through a rock garden that sits on the flanks of the ridge containing Challenger Point and Kit Carson Peak. Past this, the trail mellows again into the far reaches of this glacial valley before popping out at Willow Lake—a pristine body of water nestled at the foot of soaring peaks, snowfields, and yet another waterfall at the far end.

Back in the town of Crestone, get ready to mingle with Zen students, be serenaded by impromptu jam circles of locals playing mandolins, bongo drums, and didgeridoos, and enjoy local yak dishes and innovative brews at the relatively new Crestone Brewing Company.

TURN BY TURN DIRECTIONS:

1. Follow the trail sign for Willow Lake just past the Willow Lake/South Crestone Lake trailhead, veering right.
2. At 3.3 miles, pass a waterfall and log bridge over the creek.
3. At 4.4, pass another waterfall.
4. At Challenger/Kit Carson sign, bear right to reach Willow Lake.
5. When ready, return the same way you came.

FIND THE TRAILHEAD

As you drive into Crestone, follow signs directing you to the Willow Lake Trail. Follow Galena Avenue east out of town until it dead-ends in two miles at the trailhead. The road turns to dirt and can get rutted and bumpy, but was recently improved to be 2WD-friendly the entire way.

CRESTONE BREWING COMPANY

At its launch in 2016, Crestone Brewing Company focused on farmhouse-style saisons and other non-traditional brews (including fermented teas) made with local ingredients such as sage, lavender, and elderberry. They specialize in limited-release nanobrews (no more than three barrels per batch) to serve the offbeat community that is their home. A sampler flight is the way to go here for those who like to try bold flavors—everything from sours to IPAs to a mead-beer hybrid so strong they'll cut you off after two drinks. And don't leave without sampling the local yak burgers, vegetarian shepherd's pie, or dessert specialties.

CONTACT INFORMATION
U.S. Forest Service,
Rio Grande National Forest,
1803 W. Highway 160,
Monte Vista, CO 81144;
719-852-5941

BREWERY/RESTAURANT
Crestone Brewing Company
187 Silver Avenue,
Crestone, CO 81131
719-256-6400
Miles from trailhead: 2.3

3

SEARCH TOOL

BREWERIES

Brewery	Beer	Page
12Degree Brewing	Treachery Golden Strong Ale	164
Aspen Brewing Co.	Silver City Kettle-Soured Session Ale	76
Avalanche Brewing Co.	Pride of the West Porter	40
Bonfire Brewing	Kindler Pale Ale	116
Bristol Brewing Co.	Cheyenne Canyon Piñon Nut Brown Ale	196
Broken Compass Brewing	Coconut Porter	104
Butcherknife Brewing Co.	Bavarian Hefeweizen	120
Cerebral Brewing	Rare Trait IPA	180
Colorado Boy Pub and Brewery	Irish Red Ale	48
Crestone Brewing Co.	Emergence American Pale Ale	220
Dodgeton Creek Brewing Co.	Milk Stout	208
Dolores River Brewery	ESB	28
Eddyline Brewery	Jolly Roger Black Lager	92
Elevation Beer Co.	Apis IV Belgian Quadrupel	88
Equinox Brewing	Jonas Porter	160
Glenwood Canyon Brewpub	Dos Rios Vienna Lager	68
Gore Range Brewery	Vail Tail Pale Ale	112
Hideaway Park Brewery	Choo Choo Chai Milk Stout	132
High Alpine Brewing Co.	Sol's Espresso Stout	84
Horsefly Brewing Co.	Jazzy Razzy	52
Irwin Brewing Co.	German Pilsner	80
Kannah Creek Brewing Co.	Black's Bridge Stout	60
Lariat Lodge Brewing	Imperial Black Saison	188
Living the Dream Brewing	Helluva Caucasian Stout	176
Main Street Brewery	Schnorzenboomer Amber Doppelbock	24
Manitou Brewing Co.	High Ground IPA	200
New Terrain Brewing Co.	Cruise Ride American Cream Ale	184
Old 121 Brewhouse	Old 121 Honey Brown Ale	172
Ouray Brewery	Box Canyon Brown Ale	44
Palisade Brewing Co.	Dirty Hippie Dark American Wheat	64
Periodic Brewing	Practice Parade Scotch Ale	96
Pikes Peak Brewing	Devils Head Red Ale	192
Riff Raff Brewing Co.	El Duende Verde Chile Ale	36
Roaring Fork Beer Co.	Freestone Extra Pale Ale	72
Rock Cut Brewing Co.	Smoky Brunette Smoked Brown Ale	152
San Luis Valley Brewing Co.	Valle Especial Mexican Lager	216
Shamrock Brewing Co.	Irish Red Ale	204
South Park Brewing	Cherry Blonde Ale	100
Southern Sun Pub & Brewery	Annapurna Amber	144
Station 26 Brewing	Juicy Banger IPA	168
Steamworks Brewing Co.	Colorado Kölsch	32
Storm Peak Brewing Co.	Mad Creek Kölsch	124
Telluride Brewing Co.	Face Down Brown Ale	56
The Bakers' Brewery	Barking Dog Brown Ale	108
Three Barrel Brewing	Hop Trash IPA	212
Tommyknocker Brewery	Butt Head Bock Lager	128
Upslope Brewing Co.	Thai Style White IPA	140
Very Nice Brewing Co.	Very Nice Pale Ale	136
Wild Woods Brewery	Ponderosa Porter	148
Zwei Brewing	Zwei Pils Bavarian-style Pilsner	156

BEERS

Annapurna Amber	Southern Sun Pub & Brewery	144
Apis IV Belgian Quadrupel	Elevation Beer Co.	88
Barking Dog Brown Ale	The Bakers' Brewery	108
Bavarian Hefeweizen	Butcherknife Brewing Co.	120
Black's Bridge Stout	Kannah Creek Brewing Co.	60
Box Canyon Brown Ale	Ouray Brewery	44
Butt Head Bock Lager	Tommyknocker Brewery	128
Cherry Blonde Ale	South Park Brewing	100
Cheyenne Canyon Piñon Nut Brown Ale	Bristol Brewing Co.	196
Choo Choo Chai Milk Stout	Hideaway Park Brewery	132
Coconut Porter	Broken Compass Brewing	104
Colorado Kölsch	Steamworks Brewing Co.	32
Cruise Ride American Cream Ale	New Terrain Brewing Co.	184
Devils Head Red Ale	Pikes Peak Brewing	192
Dirty Hippie Dark American Wheat	Palisade Brewing Co.	64
Dos Rios Vienna Lager	Glenwood Canyon Brewpub	68
El Duende Verde Chile Ale	Riff Raff Brewing Co.	36
Emergence American Pale Ale	Crestone Brewing Co.	220
ESB	Dolores River Brewery	28
Face Down Brown Ale	Telluride Brewing Co.	56
Freestone Extra Pale Ale	Roaring Fork Beer Co.	72
German pilsner	Irwin Brewing Co.	80
Helluva Caucasian Stout	Living the Dream Brewing	176
High Ground IPA	Manitou Brewing Co.	200
Hop Trash IPA	Three Barrel Brewing	212
Imperial Black Saison	Lariat Lodge Brewing	188
Irish Red Ale	Colorado Boy Pub and Brewery	48
Irish Red Ale	Shamrock Brewing Co.	204
Jazzy Razzy	Horsefly Brewing Co.	52
Jolly Roger Black Lager	Eddyline Brewery	92
Jonas Porter	Equinox Brewing	160
Juicy Banger IPA	Station 26 Brewing	168
Kindler Pale Ale	Bonfire Brewing	116
Mad Creek Kölsch	Storm Peak Brewing Co.	124
Milk Stout	Dodgeton Creek Brewing Co.	208
Old 121 Honey Brown Ale	Old 121 Brewhouse	172
Ponderosa Porter	Wild Woods Brewery	148
Practice Parade Scotch Ale	Periodic Brewing	96
Pride of the West Porter	Avalanche Brewing Co.	40
Rare Trait IPA	Cerebral Brewing	180
Schnorzenboomer Amber Doppelbock	Main Street Brewery	24
Silver City Kettle-Soured Session Ale	Aspen Brewing Co.	76
Smoky Brunette Smoked Brown Ale	Rock Cut Brewing Co.	152
Sol's Espresso Stout	High Alpine Brewing Co.	84
Thai Style White IPA	Upslope Brewing Co.	140
Treachery Golden Strong Ale	12Degree Brewing	164
Vail Tail Pale Ale	Gore Range Brewery	112
Valle Especial Mexican Lager	San Luis Valley Brewing Co.	216
Very Nice Pale Ale	Very Nice Brewing Co.	136
Zwei Pils Bavarian-style Pilsner	Zwei Brewing	156

DIFFICULTIES

URBAN WALK	Duration	Loop?	
Boulder/Bobolink	30-45 min	↻	148
Louisville/Lafayette	45 min	↻	164
Denver/Rocky Mountain Arsenal	1.-1.5h	↻	168
Denver/Lakewood	45 min	↻	172
Denver/Littleton	2h	↻	176
Denver/Mile-High Loop	1-1.5h	↻	180

EASY	Duration	Loop?	
Fairplay	1.5-2h	↻	100
Winter Park	1.5h	↻	132
Monument	1-1.5h	↻	192

EASY-MODERATE	Duration	Loop?	
Pueblo	2h	↻	204
Trinidad	2-2.5h	↻	208
Cortez	2-3h	↻	24
Dolores	2h		28
Pagosa Spring	3-4h		36
Ridgway	1.5-2h	↻	48
Gunnison	2-3h	↻	84
Buena Vista	1.5-2h	↻	92
Steamboat Springs/Mad Creek	2-3h	↻	124

MODERATE	Duration	Loop?	
Telluride	1-1.5h	↻	56
Grand Junction	2-2.5h	↻	60
Aspen	2-3h		76
Leadville	2-3h	↻	96
Edwards	3-8h	↻	112
Eagle	2-3h	↻	116
Idaho Springs	1h	↻	128
Fort Collins/Reservoir Ridge	2h	↻	160
Golden	2-2.5h	↻	184
Evergreen	2.5-3.5h	↻	188
Manitou Springs	2h	↻	200
Del Norte	45 min	↻	212

MODERATE-STRENUOUS	Duration	Loop?	
Silverton	5-7h	↻	40
Montrose	1h	↻	52
Crested Butte	5-8h	↻	80
Breckenridge	4-6h		104
Steamboat Springs/Fish Creek Falls	3h		120
Nederland	4-7h	↻	136
North Boulder	1.5h	↻	140
Colorado Springs	3-4h		196
Crestone	5-7h	↻	220

STRENUOUS	Duration	Loop?	
Durango	1.5-2h	↻	32
Palisade	2-3h	↻	64
Glenwood Springs	4-5h	↻	68
Carbondale	1-1.5h	↻	72
Estes Park	5-7h		152
Fort Collins/Horsetooth Rock	3-4h	↻	156

VERY STRENUOUS	Duration	Loop?	
Ouray	4-6h	↻	44
Poncha Springs	6-8h		88
Silverthorne	5-7h		108
Boulder/Bear Peak	4-6h	↻	144
Alamosa	2h	↻	216

www.helvetiq.com